SpringerBriefs in Computer Science

Series Editors

Stan Zdonik
Peng Ning
Shashi Shekhar
Jonathan Katz
Xindong Wu
Lakhmi C. Jain
David Padua
Xuemin Shen
Borko Furht
V. S. Subrahmanian
Martial Hebert
Katsushi Ikeuchi
Bruno Siciliano

For further volumes:
http://www.springer.com/series/10028

Murat İlsever · Cem Ünsalan

Two-Dimensional Change Detection Methods

Remote Sensing Applications

Springer

Murat İlsever
Department of Computer Engineering
Yeditepe University
Ağustos Yerleşimi 26
34755 Kayisdagi
Istanbul
Turkey

Cem Ünsalan
Electrical and Electronics Engineering
Yeditepe University
Ağustos Yerleşimi 26
34755 Kayisdagi
Istanbul
Turkey

ISSN 2191-5768   ISSN 2191-5776 (electronic)
ISBN 978-1-4471-4254-6   ISBN 978-1-4471-4255-3 (eBook)
DOI 10.1007/978-1-4471-4255-3
Springer London Heidelberg New York Dordrecht

Library of Congress Control Number: 2012940953

Printed on acid-free paper

Springer is part of Springer Science+Business Media (www.springer.com)

Preface

Sequential images captured from a region may be used to detect changes there. This technique may be used in different fields such as video surveillance, medical imaging, and remote sensing. Especially in remote sensing, change detection is used in land use and cover analysis, forest or vegetation inspection, and flood monitoring. Although manual change detection is an option, the time required for it can be prohibitive. It is also highly subjective depending on the expertise of the inspector. Hence, the need for automated methods for such analysis tasks emerged. This book is about such change detection methods from satellite images. Our focus is on changes in urban regions. The layout of the book is as follows.

We start with a brief review of change detection methods specialized for remote sensing applications. While the first Earth observation satellites were equipped with 30−100 m resolution sensors; modern ones can capture images up to 0.5 m resolution. This also led to the evolution of change detection methods for satellite images. Early methods were generally pixel based. As the detail in the image increased, more sophisticated approaches emerged (such as feature based methods) for change detection.

Next, we consider pixel based change detection.We summarize well-known methods such as: image differencing, image ratioing, image regression, and change vector analysis. We introduce median filtering based background subtraction for satellite images. We also propose a novel pixel based change detection method based on fuzzy logic.

To benefit from color and multispectral information, we explore several methods such as PCA, KTT, vegetation index differencing, time dependent vegetation indices, and color invariants. Since these methods depend on a linear or a nonlinear color space transformation, we labeled them as such. Naturally, they can only be applied to the dataset having color or multispectral information.

We also considered texture based descriptors for change detection. Here, we benefit from the gray level co-occurrence matrix. We extracted four texture descriptors from it to be used for change detection. We also benefit from entropy to summarize the texture.

Different from previous approaches, we introduced a change detection framework using structure information. Here, we extract the structure in an image by edge detection, gradient magnitude based support regions, matched filtering, and local features. Graph formalism also helped us to summarize the structure in the image.

Finally, we introduced fusion of change detection methods to improve the performance. Since different change detection methods summarize the change information in different ways, they can be fused to get a better performance. Therefore, we considered the decision level fusion based on binary logic. We also developed a fusion method based on association.

We statistically evaluated the performance of the mentioned change detection methods. On a large dataset, we obtained very promising results. Especially, the change detection performance after fusion is noteworthy.

The brief summary above indicates that this book may be useful for automated change detection studies. It summarizes and evaluates the existing methods on change detection. It also proposes several novel methods for satellite image based change detection. Therefore, the interested reader may benefit from both categories to solve his or her research problems.

Istanbul, Turkey, May 2012   Murat İlsever
                                                                                               Cem Ünsalan

Acknowledgments

The authors gratefully acknowledge the financial support of The Scientific and Technological Research Council of Turkey (TUBITAK), in the framing and execution of this work through project number 110E302.

Contents

Chapter 1
Introduction

Abstract Change detection is the process of identifying differences in a region by comparing its images taken at different times. It finds applications in several fields such as video surveillance, medical imaging, and remote sensing (using satellite imagery). Several change detection applications using satellite images are in the areas of land use and cover analysis, forest or vegetation inspection, and flood monitoring. Especially for remote sensing applications, manually labeling and inspecting changes is a cumbersome task. Also, manual inspection is prone to errors and highly subjective depending on the expertise of the inspector.

Keywords Change detection · Literature review

1.1 Literature Review on Change Detection in Satellite Images

We start by giving a survey of change detection review articles in the literature. In the following sections, we explore the existing methods in detail. In this section, we benefit from these survey articles by their comparative results and a brief summary of the advantage and disadvantage of each method in the literature.

In satellite image based change detection applications, the resolution is one of the most important factors. While the first earth observation satellites (such as LANDSAT) were equipped with 30–100 meter resolution sensors; modern ones can capture images up to 0.5 meter resolution. This also led to the evolution of change detection methods for satellite images. Early methods were generally pixel based. As the detail in the image increased, more sophisticated approaches emerged (such as feature based methods) for change detection. Therefore, initial survey papers only focused on pixel based methods.

Singh [1] summarized several change detection methods such as image differencing, image regression, image ratioing, vegetation index differencing, Principal Component Analysis (PCA), post-classification comparison, and change vector analysis in terms of land cover change. Singh indicated the relationship between the land cover change and the intensity values of the satellite images as: "The basic premise in using remote sensing data for change detection is that changes in land cover must result in changes in radiance values and changes in radiance due to land cover change must be large with respect to radiance changes caused by other factors". These other factors are counted as differences in atmospheric conditions, sun angle, and soil moisture. We can add intensity variations caused by the camera to the list. These cause insignificant changes most of the times. On the other hand, we are interested in significant changes. Singh recommended the use of images taken at the same time of the year for reducing the intensity change caused by the sun angle difference and vegetation phenology change. Accurate image registration is also necessary before using satellite images for change detection. Using images without registration can lead to false alarms. In his survey paper, Singh quantitatively evaluated the change detection methods. He concluded that, image regression produced the highest change detection accuracy followed by image ratioing and image differencing. Simple techniques such as image differencing performed better than much more sophisticated transforms such as PCA.

Mas [2], in his survey paper, compared six change detection methods in terms of land cover change. He focused on a tropical area which is subject to forest clearing. Here, land cover can be classified based on the spectral reflectance of the vegetation area. Mas pointed out that, classification based on the spectral reflectance is difficult for areas where the vegetation diversity is high (such as humid tropics). Therefore, change between land cover types (presenting similar spectral signatures) is difficult to detect. As we referred previously, we can reduce the spectral change caused by the sun angle difference and vegetation phenology change by using images from the same time of the year. Mas indicated that, it is extremely difficult to obtain multi-date images taken exactly at the same time of the year, particularly in tropical regions where cloud cover is common. Therefore, he compared the performance of different change detection methods using images captured at different times of the year.

Accurate image registration is vital before using the multi-temporal images for change detection. In addition to geometric rectification, images should also be comparable in terms of radiometric characteristics. Mas referred to two ways to achieve radiometric compensation: radiometric calibration (converting images from digital number values into ground reflection values) and relative radiometric normalization between multi-temporal images. Mas reported that, relative normalization is sufficient for change detection. In relative normalization, one image is normalized using the statistical parameters of the other.

Mas grouped change detection methods under three categories as: image enhancement, multi-date data classification, and comparison of two land cover classifications. He explained each category as: "The enhancement approach involves the mathematical combination of imagery from different dates such as subtraction of bands, ratioing, image regression, and PCA. Thresholds are applied to the enhanced image to isolate the pixels that have changed. The direct multi-date classification is based on the single analysis of a combined dataset of two or more different dates, in order to identify areas of changes. The post-classification comparison is a comparative analysis of images obtained at different moments after previous independent classification". He compared the following six methods on the test area: image differencing, vegetation index differencing, selective PCA, direct multi-date classification, post-classification analysis, and combination of image enhancement and post-classification analysis. Mas reported that, post-classification comparison produced the highest accuracy. In single band analysis such as single band differencing, Landsat MSS band 2 (red) produced better results compared to Landsat MSS band 4 (infrared). Based on the same band, PCA produced better accuracy than image differencing. Superior performance of the post-classification comparison is attributed to the difficulty in classifying land cover using the spectral data. Mas indicated that, methods that are directly using the spectral data have problems in classifying land cover having similar spectral signatures. He mentioned that, the use of classification techniques avoids this problem.

Lu et al. [3] investigated a wide range of change detection techniques in their recent survey paper. They listed change detection applications which have attracted attention in the remote sensing community so far. These are: land use and land cover change, forest or vegetation change, forest mortality, defoliation and damage assessment, deforestation, regeneration and selective logging, wetland change, forest fire, landscape change, urban change, environmental change, and other applications such as crop monitoring. Lu et al. grouped change detection methods into seven categories. For our application, the most important of these are: algebra, transformation, and classification based change detection. The algebra category includes image differencing, image regression, image ratioing, vegetation index differencing and Change Vector Analysis (CVA). Lu et al. listed the advantages and disadvantages of these methods as follows. "These methods (excluding CVA) are relatively simple, straightforward, easy to implement and interpret, but these cannot provide complete matrices of change information. ... One disadvantage of the algebra category is the difficulty in selecting suitable thresholds to identify the changed areas. In this category, two aspects are critical for the change detection results: selecting suitable image bands or vegetation indices and selecting suitable thresholds to identify the changed areas". Due to the simplicity of the mentioned methods, they only provide the change and no-change information.

Lu et al. also addressed the concept of change matrix in the quotation. A change matrix covers a full range of from-to change classification. A common example includes land cover type changes such as from agricultural to urban or from forest to grassland. They considered the PCA, Kauth-Thomas (KT), Gram-Schmidt, and Chi-square transformations under the transformation category. They listed the advantages and disadvantages of these methods as follows. "One advantage of these methods is in reducing data redundancy between bands and emphasizing different information in derived components. However, they cannot provide detailed change matrices and require selection of thresholds to identify changed areas. Another disadvantage is the difficulty in interpreting and labeling the change information on the transformed images". Their classification category includes post-classification comparison, spectral-temporal combined analysis, expectation-maximization algorithm (EM) based change detection, unsupervised change detection, hybrid change detection, and artificial neural networks. The advantage of these methods is the capability of providing a change matrix. The disadvantage is the need for a qualified and large training sample set for good classification results.

1.2 Layout of the Study

In this study, we investigate several change detection methods. We group them into four categories as: pixel based, transformation based, texture based, and structure based. We explain each method in detail (with their references) in the following chapters. We also fuse the decision of these methods.

We investigate pixel based change detection methods in Chap. 2. We first focus on direct algebraic calculations such as image differencing and ratioing. Then, we consider image regression which estimates second-date image by use of linear regression. The following method is the CVA which accepts pixel values as vectors and provides change information based on vector differences. Next, we consider median filtering based background subtraction which estimates the change by subtracting multi-temporal images from a background image (e.g. an image which represents unchanged state of the observed scene in time). Finally, we develop a pixelwise fuzzy XOR operator for change detection. Among these methods, change detection using median filtering based background subtraction is a new adaptation to change detection in remote sensing. Pixelwise fuzzy XOR operator based method is the novel contribution for any change detection problem.

In Chap. 3, we investigate the transformation based change detection methods. Here, we first focus on PCA which is a common technique from the field of multivariate statistical analysis. Then, we consider the Kauth-Thomas transformation where the transformed data is directly related to the analysis of land-cover. Next, we explore vegetation index differencing and time-dependent vegetation indices (commonly used in the analysis of change in vegetation). Finally, we consider color invariants. Among these, color invariants based change detection is a novel adaptation to this field. Time-dependent vegetation indices are improved in this study.

We investigate texture based change detection methods in Chap. 4. Here, we benefit from Gray Level Co-occurrence Matrix (GLCM) features to summarize the texture information. We also benefit from the entropy of the image windows as another texture feature.

In Chap. 5, we investigate structure based change detection methods. These can be summarized as the use of edge information, gradient magnitude based support regions, matched filtering, mean shift segmentation, use of local features, graph matching with local features, and shadow information. Among these methods, use of local features, graph matching with local features and shadow information are novel contributions to change detection for remote sensing.

We propose a novel method to fuse the decision of mentioned change detection algorithms in Chap. 6. The rationale here is as follows. Each change detection method provides a change map based on its design and assumptions. Fusing their decisions may provide a better change map.

We test all mentioned change detection methods on 35 panchromatic and multispectral Ikonos satellite test image pairs. We provide the quantitative comparison results on these images as well as the strengths and weaknesses of each method in Chap. 7. Finally, in Chap. 8, we summarize the conclusions that we have reached in this study.

References

1. Singh, A.: Review article: digital change detection techniques using remotely-sensed data. Int. J. Remote Sens. **10**(6), 989–1003 (1989)
2. Mas, J.F.: Monitoring land-cover changes: a comparison of change detection techniques. Int. J. Remote Sens. **20**(1), 139–152 (1999)
3. Lu, D., Mausel, P., Brondizio, E., Moran, E.: Change detection techniques. Int. J. Remote Sens. **25**(12), 2365–2401 (2004)

Chapter 2
Pixel-Based Change Detection Methods

Abstract In this chapter, we consider pixel-based change detection methods. First, we provide well-known methods in the literature. Then, we propose two novel pixel-based change detection methods.

Keywords Pixel-based change detection · Image differencing · Automated thresholding · Percentile · Otsu's method · Kapur's algorithm · Image rationing · Image regression · Least-squares · Change vector analysis (CVA) · Median filtering · Background formation · Fuzzy logic · Fuzzy xor

2.1 Image Differencing

In this technique, images of the same area, obtained from times t_1 and t_2, are subtracted pixelwise. Mathematically, the difference image is

$$I_d(x, y) = I_1(x, y) - I_2(x, y), \tag{2.1}$$

where I_1 and I_2 are the images obtained from t_1 and t_2, (x, y) are the coordinates of the pixels. The resulting image, I_d, represents the intensity difference of I_1 from I_2. This technique works only if images are registered.

To interpret the difference image, we need to recall the quotation from Singh [1]: "The basic premise in using remote sensing data for change detection is that changes in land cover must result in changes in radiance values and changes in radiance due to land cover change must be large with respect to radiance changes caused by other factors." Based on this principle, we can expect that intensity differences due to land cover change resides at the tails of the difference distribution of the image. Assuming that changes due to land cover are less than changes by other factors, we expect that most of the difference is distributed around the mean. We can illustrate the difference distribution as in Fig. 2.1.

M. İlsever and C. Ünsalan, *Two-Dimensional Change Detection Methods*,   7
SpringerBriefs in Computer Science, DOI: 10.1007/978-1-4471-4255-3_2,
© Cem Ünsalan 2012

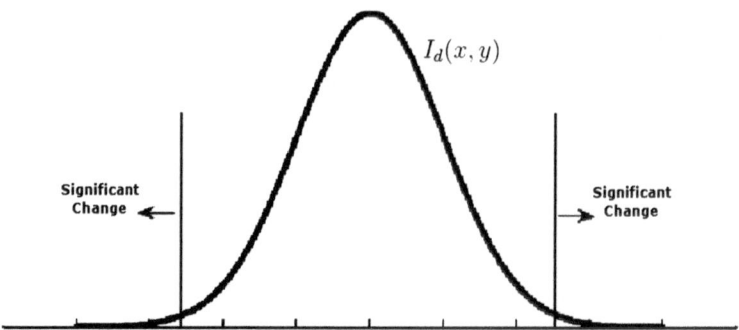

$I_d(x, y)$

Significant
Change

Significant
Change

Fig. 2.1 Distribution of a difference function. Significant changes are expected at the tails of the distribution

For a zero mean difference distribution, we can normalize I_2 as

$$\tilde{I}_2(x, y) = \frac{\sigma_1}{\sigma_2}(I_2(x, y) - \mu_2) + \mu_1, \tag{2.2}$$

where $\tilde{I}_2$ is the normalized form of I_2. μ_1, σ_1 and μ_2, σ_2 are the mean and the standard deviation of I_1 and I_2, respectively. After normalization, the mean and standard deviation of the two images are equalized. Hence, the difference image will have zero mean. Now, we can update Eqn. 2.1 as

$$I_d(x, y) = |I_1(x, y) - \tilde{I}_2(x, y)|. \tag{2.3}$$

To detect the change, we can apply simple thresholding to $I_d(x, y)$ as

$$T(x, y) = \begin{cases} 1, & I_d(x, y) \geq \tau \\ 0, & \text{otherwise,} \end{cases} \tag{2.4}$$

where the threshold τ is often determined empirically.

Since the threshold value in Eqn. 2.4 is important, various automated threshold selection algorithms are proposed. Most of the time, the performance of these algorithms is scene dependent due to the assumptions they are based on. Rosin and Ioannidis [2] investigated the performance of several automated thresholding algorithms using a large set of difference images calculated from an automatically created ground truth database. They give results based on several measures for a complete evaluation. In this study, we benefit from three different threshold selection methods. These are percentile thresholding, Otsu's method [3] and Kapur's algorithm [4].

We will briefly explain these thresholding methods next. To this end, we make some assumptions about I_d as follows. I_d is a grayscale image which is represented by N_g gray levels, $\{1, 2, \ldots, N_g\}$. The number of pixels at level i is denoted by n_i and the total number of pixels is N.

The first thresholding method is based on the percentile [5]. It is a statistics of ordinal scale data. Assume that A is a sorted array of pixel values of I_d in ascending order. Rank of the Pth percentile of I_d is given by

$$R = \text{ceil}\left(\frac{P}{100} \times N\right) \tag{2.5}$$

where the ceil function rounds its argument to the nearest greater integer. Pth percentile is found by indexing A using that rank.

The second thresholding method is proposed by Otsu. It uses measures of class separability in finding an optimal threshold value. Relative frequencies of pixel values at level i are given by

$$p_i = \frac{n_i}{N}, \quad p_i \geq 0, \quad \sum_{i=1}^{N_g} p_i = 1. \tag{2.6}$$

A threshold value at gray level k divides the histogram of I_d into two classes. Each class has its own probability of occurrence (total probability of its samples) and own mean value. Evaluation function of the Otsu's method is the between-class variance given by

$$\sigma_b^2 = \frac{[\mu_{I_d}\omega(k) - \mu_\omega]^2}{\omega(k)[1 - \omega(k)]}, \tag{2.7}$$

where μ_{I_d} is the mean of I_d; $\omega(k)$ is the probability of the class which includes gray levels up to k and μ_ω is the mean of the class ω. The optimal threshold value k^* maximizes

$$\sigma_b^2(k^*) = \max_{1 \leq k \leq N_g} \sigma_b^2(k). \tag{2.8}$$

The last thresholding method we use is Kapur's algorithm. Similar to Otsu's method, it divides the image histogram into two classes. It then utilizes the sum of the entropy of these two classes as an evaluation function. The value which maximizes this sum is taken as the optimal threshold value. For two classes A and B, Shannon entropy of these classes are defined as

$$H(A) = -\sum_{i=1}^{k} \frac{p_i}{\omega(k)} \ln \frac{p_i}{\omega(k)}, \tag{2.9}$$

$$H(B) = -\sum_{i=k+1}^{N_g} \frac{p_i}{[1 - \omega(k)]} \ln \frac{p_i}{[1 - \omega(k)]}, \tag{2.10}$$

where the histogram is divided at gray level k. The optimal threshold value k^* maximizes the sum $\phi(k) = H(A) + H(B)$ such that

Fig. 2.2 Images taken at two different times from a developing region of Adana

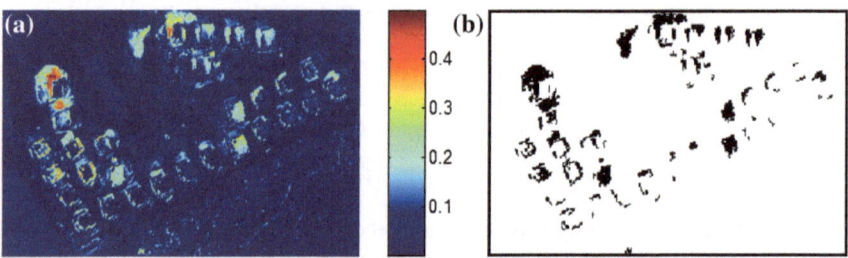

Fig. 2.3 Image differencing applied to the Adana image set. **a** The difference image **b** Thresholded version

$$\phi(k^*) = \max_{1 \leq k \leq N_g} \phi(k). \tag{2.11}$$

To explain different change detection methods, we pick the Adana test image set given in Fig. 2.2. The two images, taken in different times, in this set represent a region with construction activity. These images are registered. Therefore, they can be used for pixelwise change detection methods. The difference between these two images is clearly seen. We will use this image set in the following sections also.

The difference image obtained from the Adana image set is as in Fig. 2.3a. This image is color coded with the color scale given next to it. We also provide the thresholding result in Fig. 2.3b. In thresholding, we benefit from Kapur's method. As can be seen, the thresholded image provides sufficient information about the changed regions in the image.

Griffiths [6] used image differencing for detecting the change in urban areas. He used Landsat TM data (with 30 m resolution), SPOT XS multispectral data (with 20 m resolution), and SPOT panchromatic data (with 10 m resolution) in his study. He proposed using an urban mask to find changes in urban areas using image differencing. Griffiths indicated that the mixture of buildings, streets, and small gardens in urban areas produce a highly textured appearance compared with the much smoother texture of arable fields. He used a standard deviation filter to quantify the texture. The urban mask is multiplied by the difference image to eliminate non-urban areas. Furthermore, he refined the results based on a previous study. In this technique, changes that occur far from the urban areas are assumed to be non-urban change. This is because new urban development generally occurs at the periphery of existing urban areas. Griffiths presented his results for each technique by visual interpretation.

Saksa et al. [7] used image differencing for detecting clear cut areas in boreal forests. They tested three methods using Landsat satellite imagery and aerial photographs as: pixel-by-pixel differencing and segmentation, pixel block-level differencing and thresholding, pre-segmentation and unsupervised classification. In the first method, they found the difference image. Then, they used a segmentation algorithm to delineate the clear cut areas. In the second method, they included neighboring pixels into the calculation of the difference image. Therefore, negative effects of misregistration are reduced in the resulting image. In the third method, they first segmented the images. Then, they obtained a segment-level image difference. They labeled clear cut areas by using an unsupervised classification algorithm. Saksa et al. concluded that, predelineated segments or pixel blocks should be used for image differencing in order to decrease the amount of misinterpreted small areas.

Lu et al. [8] compared 10 binary change detection methods to detect land cover change in Amazon tropical regions. They used Landsat TM (Thematic Mapper) data in their study. In addition to band differencing, they tested a modified version of image differencing where pixels are accepted as changed when majority of the bands indicate change. For six-band Landsat TM data, if four of the bands indicate change then the pixel value is labeled as changed. They reported that the difference of Landsat TM band 5, modified image differencing, and principal component differencing produced best results.

2.2 Image Rationing

Similar to image differencing, images are compared pixelwise in this method. Therefore, images must be registered beforehand. The ratio image, used in this method, is calculated by

$$I_r(x, y) = \frac{I_1(x, y)}{\tilde{I}_2(x, y)}. \tag{2.12}$$

In Eqn. 2.12, the I_r image takes values in the range $[0, \infty)$. If the intensity values are equal, it takes the value 1. To normalize the value of I_r, we can benefit from the arctangent function as

$$I_r(x, y) = \arctan\left(\frac{I_1(x, y)}{\tilde{I}_2(x, y)}\right) - \frac{\pi}{4}. \tag{2.13}$$

Now, ratio image takes values in the range $[-\pi/4, \pi/4]$. To threshold I_r, we can benefit from the same methods as we did in the previous section. In Fig. 2.4a, we provide the I_r image obtained from the Adana test image set. We provide the thresholded version of this image in Fig. 2.4b. As in the previous section, we used

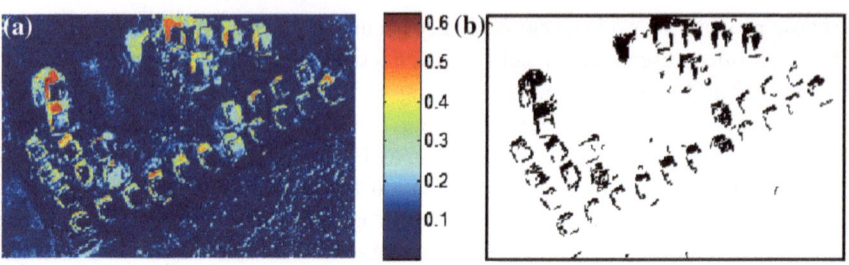

Fig. 2.4 Image ratio applied to the Adana image set. **a** The ratio image **b** Thresholded Version

Kapur's method to obtain the optimal threshold value. As can be seen, the thresholded image provides sufficient information on the changed regions in the image.

2.3 Image Regression

In image regression, the I_2 image (obtained from t_2) is assumed to be a linear function of the I_1 image (obtained from t_1). Under this assumption, we can find an estimate of I_2 by using least-squares regression as

$$\hat{I}_2(x, y) = aI_1(x, y) + b. \tag{2.14}$$

To estimate the parameters a and b, we define the squared error between the measured data and predicted data (for each pixel) as

$$e^2 = (I_2(x, y) - \hat{I}_2(x, y))^2 = (I_2(x, y) - aI_1(x, y) - b)^2. \tag{2.15}$$

The sum of the squared error becomes

$$S = \sum_{n=1}^{N} e^2 = \sum_{n=1}^{N} (I_2(x_n, y_n) - aI_1(x_n, y_n) - b)^2. \tag{2.16}$$

Here, we assume that we have N observations. We want to find the parameters a and b to minimize the sum of the squared error S. Therefore, we first calculate the partial derivatives of S with respect to a and b as

$$\frac{\partial S}{\partial b} = -2 \sum_{n=1}^{N} (I_2(x_n, y_n) - aI_1(x_n, y_n) - b), \tag{2.17}$$

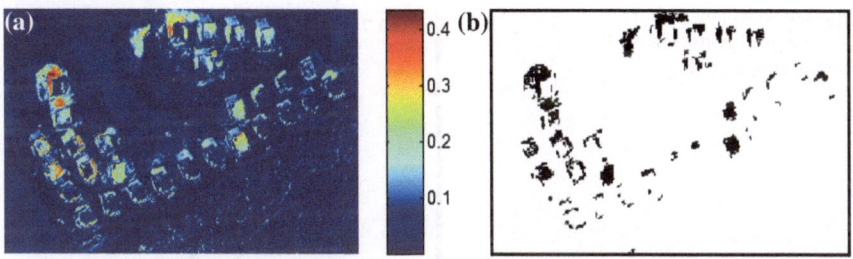

Fig. 2.5 Image difference after regression applied to the Adana image set. **a** The difference image **b** Threshold Version

$$\frac{\partial S}{\partial a} = -2 \sum_{n=1}^{N} [(I_2(x_n, y_n) - aI_1(x_n, y_n) - b)I_1(x_n, y_n)]. \qquad (2.18)$$

By equating Eqn. 2.17 and 2.18 to zero, we obtain two equations with two unknowns as

$$0 = \sum_{n=1}^{N} I_2(x_n, y_n) - \sum_{n=1}^{N} aI_1(x_n, y_n) - \sum_{n=1}^{N} b, \qquad (2.19)$$

$$0 = \sum_{n=1}^{N} I_2(x_n, y_n)I_1(x_n, y_n) - \sum_{n=1}^{N} aI_1(x_n, y_n)^2 - \sum_{n=1}^{N} bI_1(x_n, y_n). \qquad (2.20)$$

Solving these equations, we obtain

$$a = \frac{n \sum_{n=1}^{N} I_2(x_n, y_n)I_1(x_n, y_n) - \sum_{n=1}^{N} I_2(x_n, y_n) \sum_{n=1}^{N} I_1(x_n, y_n)}{n \sum_{n=1}^{N} I_1(x_n, y_n)^2 - (\sum_{n=1}^{N} I_1(x_n, y_n))^2}, \qquad (2.21)$$

$$b = \frac{\sum_{n=1}^{N} I_2(x_n, y_n) - a \sum_{n=1}^{N} I_1(x_n, y_n)}{N}. \qquad (2.22)$$

We manually picked the observations (for $n = 1, \ldots, N$) from I_1 and I_2 (from the unchanged areas). When we subtract I_2 from $\hat{I}_2$ as $I_d(x, y) = I_2(x, y) - \hat{I}_2(x, y)$, we expect to find changes originating from land cover. When we apply this method to the normalized I_2 ($\tilde{I}_2$), we further eliminate the insignificant changes that still remain after normalization. Consequently, this technique gives slightly better performance compared to image differencing.

We provide the difference image obtained by image regression using our Adana image test set in Fig. 2.5a. We provide the thresholded version of this image in Fig. 2.5b. As in the previous sections, we benefit from Kapur's method in threshold selection. As can be seen, the change map obtained is similar to image differencing.

Fig. 2.6 Unchanged and
changed pixel vectors in a 2-D
spectral space

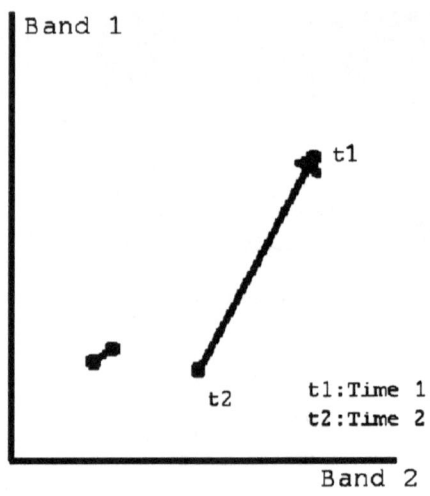

2.4 Change Vector Analysis

Change Vector Analysis (CVA) is a technique where multiple image bands can be analyzed simultaneously. As its name suggests, CVA does not only function as a change detection method, but also helps analyzing and classifying the change. In CVA, pixel values are vectors of spectral bands. Change vectors (CV) are calculated by subtracting vectors pixelwise as in image differencing. The magnitude and direction of the change vectors are used for change analysis. In Fig. 2.6, a changed pixel and an unchanged pixel are given in a two-band spectral space.

The change vector magnitude can indicate the degree of change. Thus, it can be used for change and no-change classification. Under ideal conditions, such as perfect image registration and normalization, unchanged pixel magnitudes must be equal to zero. However, this is not the case in practice. Therefore, thresholding must be applied to the change magnitude. While the change vector magnitude behaves like a multi-band version of the image differencing, the change direction gives us additional information about the change type. This is often more valuable than the amount of change, since in most applications we are interested in a specific change type.

In practice, the number of CV directions are uncountable. Therefore, it is necessary to quantize the CV space and assign directions accordingly. A simple quantization of CV directions can be achieved by dividing the space by its main axes. In Fig. 2.7, a 2D CV space is quantized into four subsets (quadrants) by the axis of band 1 and band 2. For three band images, subsets can be octants. CVs can be assigned to subsets via signs of their components.

As mentioned earlier, CV directions can be used in classifying the change. By using subsets, we can determine 2^n classes of change for an n dimensional space. CVA

Fig. 2.7 Change vector space
is divided into four subsets

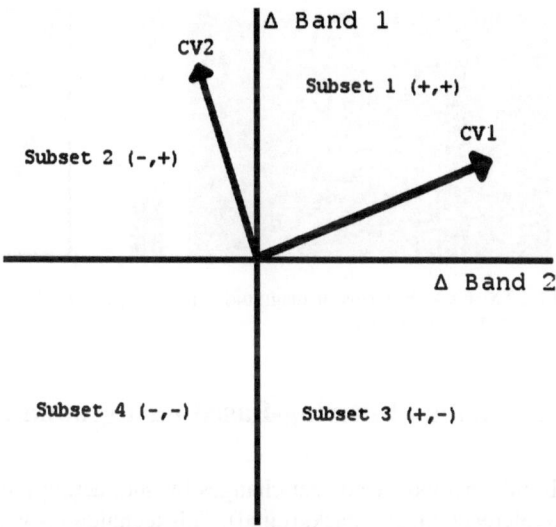

can also be applied to transformed data such as Kauth-Thomas Transformation (KTT) (explained in Sect. 3.2) rather than to raw data. In the KTT space, a simultaneous increase in the greenness feature and decrease in the brightness feature indicates gain of vegetation. Therefore, in the change vector space of KTT bands, we can assign this change class (change toward vegetation) to the subsets where greenness is positive and brightness is negative.

CVA is used first by Malila [9] for change detection. He used the KTT with CVA and reported results for change of forestation. He used change directions to distinguish changes due to harvesting and regrowth. Johnson et al. [10] provided a comprehensive investigation of CVA. They provided the details to the implementation of CVA after a functional description. They reported that, CVA can be used in applications which require a full-dimensional data processing and analysis technique. They also found CVA to be useful for applications in which: the changes of interest and their spectral manifestation are not well-known a priori; changes of interest are known or thought to have high spectral variability; changes in both land cover type and condition may be of interest.

In Fig. 2.8a, we provide the change vector magnitude image which can be used for change and no-change classification. In Fig. 2.8b the change vector magnitude image is thresholded by Kapur's algorithm. This thresholded image also provides sufficient information on changed regions.

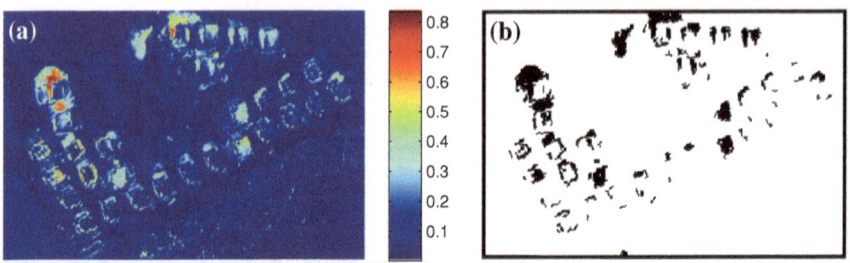

Fig. 2.8 CVA (in terms of magnitude value) applied to the Adana image set. **a** The magnitude value **b** Threshold Version

2.5 Median Filtering-Based Background Formation

In this method, we detect changes by subtracting the multi-temporal images from a reference image (background). This technique is widely used in video processing for detecting and analyzing motion. Although well known in video signal processing community, this is the first time median filtering is used for change detection in satellite images.

For this method, several background formation techniques are proposed in the literature. In this study, we use temporal median filtering for background formation. It is defined as

$$I_{bg}(x, y) = \text{Med}(I_1(x, y), I_2(x, y), \dots, I_N(x, y)), \qquad (2.23)$$

where I_{bg} is the background image and $(I_1, I_2, \dots, I_N)$ are the images from times $(t_1, t_2, \dots, t_N)$.

Parameters of the median operator are pixel values from the same spatial location (x, y). Median is calculated by choosing the middle element of the sorted array of its parameters. This leads to the removal of outliers (impulsive or salt and pepper noise) from the input pixel set. This characteristic of the median filter helps us to find a pixel value for every spatial location which is equal or approximately equal to the majority of the elements of the temporal pixel set. We expect to find the change by subtracting each image from the background, thus enchanting deviations from the median value.

We provide our multi-temporal images for background formation in Fig. 2.9. We also provide the median filtering result in Fig. 2.10.

Using the extracted background image, we obtain four difference images as in Fig. 2.11. We also provide the thresholded version of these images in Fig. 2.12. As in the previous sections, we used Kapur's algorithm in finding the threshold value. As can be seen in Fig. 2.12, majority of the significant changes are gathered in the first and last images in the series. These correspond to the two extremes of the time interval.

Fig. 2.9 Multi-temporal images used in background formation

Fig. 2.10 Background image
generated by median filtering

2.6 Pixelwise Fuzzy XOR Operator

The last method for pixelwise change detection is a novel contribution to the community. In this method, the binary XOR operation is taken as a benchmark. Its fuzzy version is used for change detection. Our rationale here is as follows. Assume that we have two binary images (composed of only ones and zeros) and we want to detect the changed pixels in these. Each pixel $p(x, y)$ in a binary image B is valued according to a characteristic function β_B, which could also be called as the "whiteness" function defined as

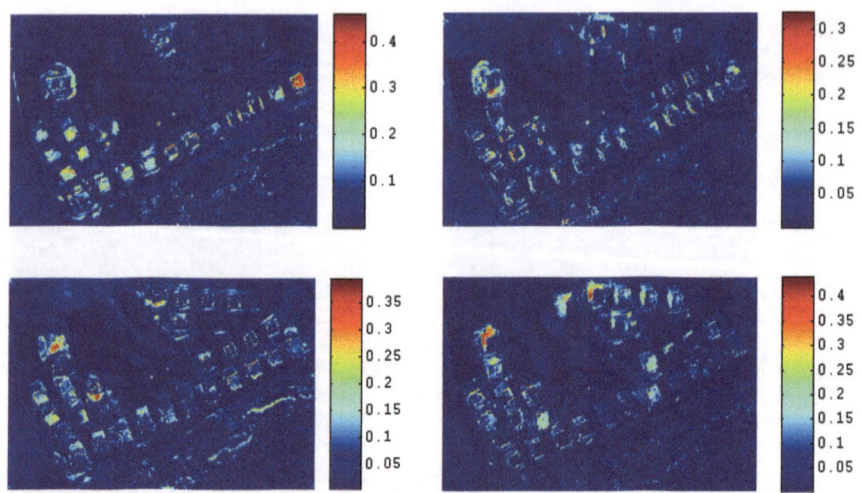

Fig. 2.11 Difference images for each sample generated by subtracting each sample from the background image

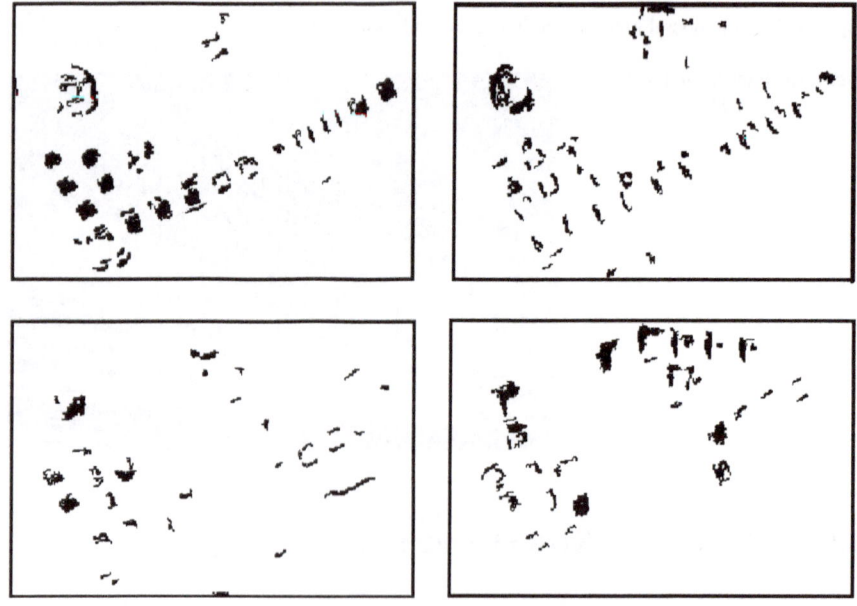

Fig. 2.12 Difference images are thresholded by Kapur's algorithm

$$p(x, y) = \beta_B(x, y) = \begin{cases} 1, & \text{if } B(x, y) \text{ is white} \\ 0, & \text{otherwise.} \end{cases} \tag{2.24}$$

Between two pixels p_1 and p_2, at the same (x, y) coordinates of the two binary images B_1 and B_2, the existence of a change can only mean that either "p_1 is white and p_2 is not" or "p_1 is not white and p_2 is." This directly implies the XOR operation in binary logic. Hence the obvious solution to the change detection problem is XOR-ing the two binary images pixelwise as

$$C(x, y) = B_1(x, y) \oplus B_2(x, y). \tag{2.25}$$

This operation gives '0' for pixels having the same value in both images, and gives '1' for pixels having different values. Therefore, white pixels in the resulting binary image $C(x, y)$ represent the changed regions.

Unfortunately, this method cannot be applied to panchromatic or multispectral satellite imagery (having pixel values in a certain range). In order to perform a similar method on satellite imagery, we propose a fuzzy representation for these. We also benefit from the combination of fuzzy and crisp (binary) operations.

Panchromatic images are composed of pixels with values $p(x, y)$ in a certain range. Normalizing these values and mapping them to the range $[0, 1]$ effectively translates the image into a fuzzy set, whose elements (pixels) have membership grades in proportion to their "whiteness." The membership grade $g(x, y)$ of each pixel $p(x, y)$ in the grayscale image G is thus defined by the fuzzy membership function μ_G as

$$g(x, y) = \mu_G(x, y) = \begin{cases} 1.00, & \text{if } G(x, y) \text{ is pure white} \\ \dots & \dots \\ 0.50, & \text{if } G(x, y) \text{ is gray} \\ \dots & \dots \\ 0.00, & \text{if } G(x, y) \text{ is pure black.} \end{cases} \tag{2.26}$$

Comparison of two binary images involves the crisp question "Are these two pixels different?." Whereas a fuzzy comparison of two panchromatic images involves the fuzzy question "How different are these two pixels?." Also the question of "Above what amount of difference shall the two pixels be labeled as changed?." The amount of difference between gray level values in the image domain directly corresponds to the difference between the degrees of membership in the fuzzy domain. For this particular application, the fuzzy complement (NOT) operation, defined as

$$\bar{g}(x, y) = \mu_{\bar{G}}(x, y) = 1 - g(x, y) \tag{2.27}$$

and the algebraic representation of the fuzzy intersection (AND) operation, defined as the multiplication of membership functions

$$\mu_{G_1 \cap G_2} = \mu_{G_1}(x, y)\mu_{G_2}(x, y) = g_1(x, y)g_2(x, y) \tag{2.28}$$

were used to obtain a fuzzy difference metric [11].

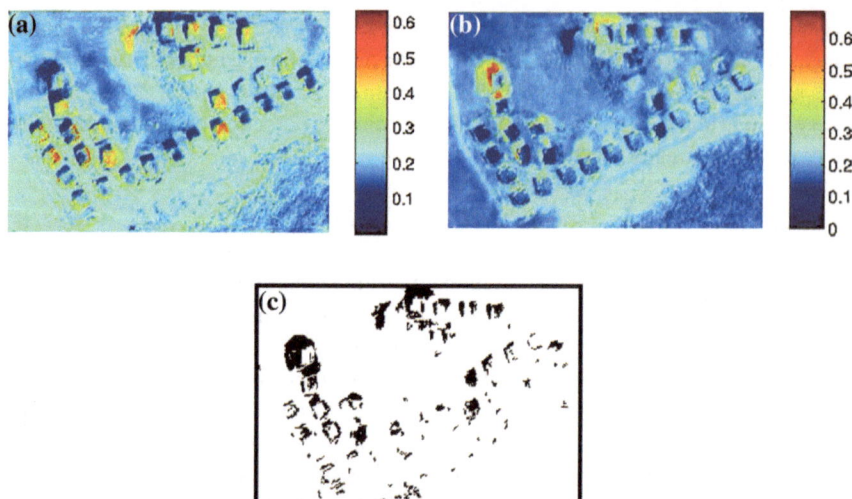

Fig. 2.13 Fuzzy XOR applied to the Adana image set. **a** Fuzzy AND $(g_1(x, y)\bar{g}_2(x, y))$ **b** Fuzzy AND $(\bar{g}_1(x, y)g_2(x, y))$ **c** Thresholded version

In a manner similar to the binary case, the measure of change between two pixels p_1 and p_2 is given by the degree of truth of the following statement: either "p_1 is lighter AND p_2 is darker" OR "p_1 is darker AND p_2 is lighter"; which can be rephrased as, either "p_1 has a high membership grade AND p_2 has a low membership grade" OR "p_1 has a low membership grade AND p_2 has a high membership grade."

Considering that "having a low membership grade" is the opposite of "having a high membership grade," the former statement's degree of truth is the complement of the latter's, and the degree of truth in "having a high membership grade" is equivalent to the membership grade $g(x, y)$ itself. Consequently, the above fuzzy rule can be formulated as

$$C(x, y) = \mu_{(G_1 \cap \bar{G}_2) \cup (\bar{G}_1 \cap G_2)}(x, y) = (g_1(x, y)\bar{g}_2(x, y)) \cup (\bar{g}_1(x, y)g_2(x, y)). \tag{2.29}$$

The fuzzy value $C(x, y)$ represents the measure of change between two images at the coordinate (x, y). The decision of a significant change can be made by means of applying an appropriate threshold and converting $C(x, y)$ to a crisp YES/NO value. Experiments have shown that, the results from the two fuzzy AND operations are distributed in a way that automatically indicates an appropriate threshold for defuzzification. More explicitly, threshold values are obtained for both fuzzy AND operations from $\tau = \text{argmax}(H_a) + 2\sigma_a$. Here, H_a is the histogram of the corresponding fuzzy AND operation and σ_a is the standard deviation of the corresponding fuzzy AND operation. In fact, applying this threshold and converting the fuzzy AND results to a crisp binary value, and then combining them with the binary OR operator yielded better results in detecting changed regions in satellite images. Therefore, the proposed method was eventually established as an ensemble of both fuzzy and binary logic operations.

We provide the images obtained by fuzzy AND operations using our Adana image set in Fig. 2.13a and Fig. 2.13b. We provide the thresholded version after finding the $C(x, y)$ in Fig. 2.13c. As can be seen, the resulting image shows the changed region fairly well.

References

1. Singh, A.: Review article: Digital change detection techniques using remotely-sensed data. Int. J. Remote Sens. **10**(6), 989–1003 (1989)
2. Rosin, P.L., Ioannidis, E.: Evaluation of global image thresholding for change detection. Pattern Recognition Lett. **24**(14), 2345–2356 (2003)
3. Otsu, N.: A threshold selection method from gray-level histograms. IEEE Trans. Syst. Man Cybern. **9**(1), 62–66 (1979)
4. Kapur, J.N., Sahoo, P.K., Wong, A.K.C.: A new method for gray-level picture thresholding using the entropy of the histogram. Comput. Vis. Graph. Image Process. **29**(3), 273–285 (1985)
5. Devore, J.: Probability and Statistics for Engineering and Sciences. 6 edn. Thompson (2004)
6. Griffiths, G.H.: Monitoring urban change from Landsat TM and Spot satellite imagery by image differencing. In: Proceedings of the 1988 International Geoscience and Remote Sensing Symposium, vol. 1, (1988)
7. Saksa, T., Uuttera, J., Kolstrom, T., Lehikoinen, M., Pekkarinen, A., Sarvi, V.: Clear-cut detection in boreal forest aided by remote sensing. Scandinavian J. For. Res. **18**(6), 537–546 (2003)
8. Lu, D., Mausel, P., Batistella, M., Moran, E.: Land-cover binary change detection methods for use in the moist tropical region of the Amazon: A comparative study. Int. J. Remote Sens. **26**(1), 101–114 (2005)
9. Malila, W.A.: Change vector analysis: An approach for detecting forest changes with Landsat. In: LARS Symposia, p. 385 (1980)
10. Johnson, R.D., Kasischke, E.S.: Change vector analysis: A technique for the multispectral monitoring of land cover and condition. Int. J. Remote Sens. **19**(3), 411–426 (1998)
11. Klir, G.J., Yuan, B.: Fuzzy Sets and Fuzzy Logic Theory and Applications. Prentice Hall, New York (1995)

Chapter 3
Transformation-Based Change Detection Methods

Abstract This chapter deals with change detection methods based on color or multispectral space transformations. They are based on Principal Component Analysis (PCA), Kauth-Thomas transformation, vegetation indices, and color invariants.

Keywords Transformation-based change detection · Principal component analysis (PCA) · Eigenvectors · Eigenvalues · Kauth-Thomas transformation · Tasseled cap transformation · Gram-Schmid orthogonalization · Ratio vegetation index (RVI) · Normalized difference vegetation index (NDVI) · Transformed vegetation index (TVI) · Soil adjusted vegetation index (SAVI) · Modified soil adjusted vegetation index (MSAVI) · Time-dependent vegetation indices (TDVI) · Color invariants

3.1 Principal Component Analysis

Principal Component Analysis (PCA) is the transformation of the multivariate data to a new set of components where data variation can be expressed by a first few components. PCA achieves this by removing the redundancy in the data set. This redundancy is quantified by the correlation of the variables. Hence, PCA transforms a correlated set of data to an uncorrelated set.

In terms of linear algebra, what PCA does is a basis rotation. This can be defined in an algorithmic manner as follows. Variance of the projections onto the first basis vector (e_1) takes its maximum after the rotation. Under the assumption that e_1 is fixed (e.g rotation axis is e_1), variance of the projections onto the second basis vector (e_2) takes its possible maximum after the rotation. Variance of the projections onto the basis vector e_m takes its possible maximum under the assumption that vectors from the previous steps ($e_1, e_2, \ldots, e_{m-1}$) are fixed. Data is redefined under a new basis ($e_1, e_2, \ldots, e_n$).

PCA is algebraically defined as follows. The sample covariance of N observations of K variables ($X_1, X_2, \ldots, X_K$) is the $K \times K$ matrix $C_x = [c_{jk}]$ with entries

M. İlsever and C. Ünsalan, *Two-Dimensional Change Detection Methods*,
SpringerBriefs in Computer Science, DOI: 10.1007/978-1-4471-4255-3_3,
© Cem Ünsalan 2012

$$c_{jk} = \frac{1}{N-1} \sum_{i=1}^{N} (x_{ij} - \mu_j)(x_{ik} - \mu_k), \qquad (3.1)$$

where x_{ij} is the ith observation of the jth variable. μ_j and μ_k are the mean of jth and kth variables, respectively. Based on these, the PCA transformation can be defined as $\mathbf{Y} = \mathbf{XU}$ where $\mathbf{U}$ is the $K \times K$ rotation matrix whose columns are the eigenvectors of C_x. $\mathbf{X}$ is the $N \times K$ data matrix whose columns and rows represent the variables and observations respectively. Columns of $\mathbf{Y}$ are Principal Components (PCs). Correlation of the variables to the PCs is a special measure. This is called as principal component loadings. The principal component loadings indicate how much variance in each of the variables is accounted for by the PCs.

Application of the PCA to change detection requires the analysis of the PC loadings. There exist two approaches to analyze the multi-temporal images in the context of change detection. The first approach is called *separate rotation*. In this approach, the PCA is applied to multi-band images separately. Then, any of the change detection techniques such as image differencing is applied to the PCs. The second approach is called *merged rotation*. In this approach, data from the bi-temporal images are merged into one set and PCA is applied to it. PCs which account for the change are selected via analysis of the PC loadings. These PCs have negative correlation to the bi-temporal data.

Fung and LeDrew [1] applied PCA to land cover change detection. They calculated the eigenvectors from the correlation and covariance matrices. Then, they compared the change detection performance when the PCA is applied using each eigenvector. They reported that PCA with eigenvectors calculated from the correlation matrix gives better results. They first reported results from the separate rotation of the multi-temporal data. They indicated that a careful examination of the PC loadings is necessary before applying change detection to PCs obtained after separate rotation. Second, they analyzed the results from the merged rotation. They listed the PCs which are responsible for the change in terms of brightness and greenness. They reported that, PC loadings from the correlation matrix are better aligned compared to PC loadings from the covariance matrix. In applying the PCA to our test images, we observed that the separate rotation approach gives better results compared to the merged rotation approach.

We used multispectral images in our PCA application. Therefore, we benefit from the near infrared band as well as visible red, green, and blue bands. Since multispectral images have lower resolution compared to their panchromatic counterparts, obtained results are not as rich as panchromatic images in terms of visual interpretation. Still PCA gives good results compared to the pixel-based methods. In Fig. 3.1, we provide the differences of principal components for the Adana image set. As in the previous sections, we benefit from Kapur's method in threshold selection. As can be seen, the difference of the third principal components emphasized changes fairly well.

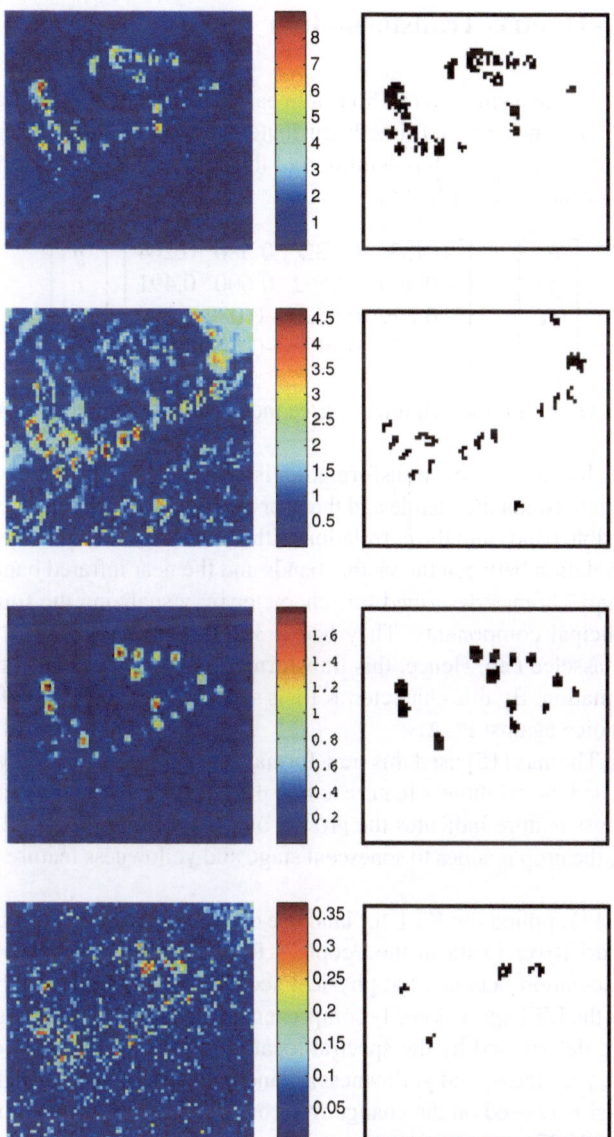

Fig. 3.1 Difference of principal components for the Adana image set. *First column* first PC, second PC, third PC3, fourth PC; *second column* their thresholded versions

3.2 Kauth-Thomas Transformation

Kauth-Thomas transformation (KTT) is a linear transformation from the multispectral data space to a new space (directly attributed to analyze the land cover) [15]. In fact, similar to PCA, KTT is a redefinition of the data. Different from PCA, KTT is a fixed transformation described as

$$
\begin{bmatrix} br \\ gr \\ ye \\ ns \end{bmatrix} = \begin{bmatrix} 0.433 & 0.632 & 0.586 & 0.264 \\ -0.290 & -0.562 & 0.600 & 0.491 \\ -0.829 & 0.522 & -0.039 & 0.194 \\ 0.223 & 0.012 & -0.543 & 0.810 \end{bmatrix} \begin{bmatrix} g \\ r \\ n_1 \\ n_2 \end{bmatrix}, \tag{3.2}
$$

where br, gr, ye, ns are the brightness, greenness, yellowness, and nonsuch values, respectively.

The fixed character of this transformation is explained by the invariant nature of the correlation between the visible and the near infrared bands. While the correlation within the visible bands and the correlation within the near infrared bands are always high, the correlation between the visible bands and the near infrared bands is always low. Kauth and Thomas described this character by visualizing the four band data via their principal components. They had a 3D representation of the data which resembles a tasseled cap. Hence, this transformation is also known as the Tasseled Cap transformation. By this character, KTT is scene independent and often referred as a better choice against PCA.

Kauth and Thomas [15] used this transformation to describe the lifecycle of croplands. They used the brightness feature to find the soil where crop grows on. Increase in the greenness feature indicates the growth of crop until it matures. At the end of the lifecycle, the crop reaches to senescent stage and yellowness feature increases in parallel.

Seto et al. [3] applied the KTT for land use change detection in a fast developing area, The Pearl River Delta in the People?s Republic of China. They referred to the direct association between the physical scene attributes and KTT bands. They observed that the KTT space is easily comprehensible. Land cover types such as forest and urban are determined by the spectral locations in the KTT space (e.g. amount of brightness, greenness, and yellowness). Land use change, such as agricultural to urban, is classified based on the change from one land cover type to another.

In applying KTT to our test images, we first need a transformation matrix convenient for Ikonos images. As defined earlier, KTT is applied to the data from green, red, and two near infrared bands. However, in Ikonos images we have only one near infrared band. In order to use the KTT matrix with green, red, and one near infrared band, we first remove the fourth row and fourth column of the matrix. Since resulting matrix is not orthogonal, we applied Gram-Schmidt orthogonalization [4] and obtain

$$
\begin{bmatrix} br \\ gr \\ ye \end{bmatrix} = \begin{bmatrix} 0.449 & 0.655 & 0.608 \\ -0.267 & -0.551 & 0.791 \\ -0.853 & 0.517 & 0.072 \end{bmatrix} \begin{bmatrix} g \\ r \\ n \end{bmatrix}. \tag{3.3}
$$

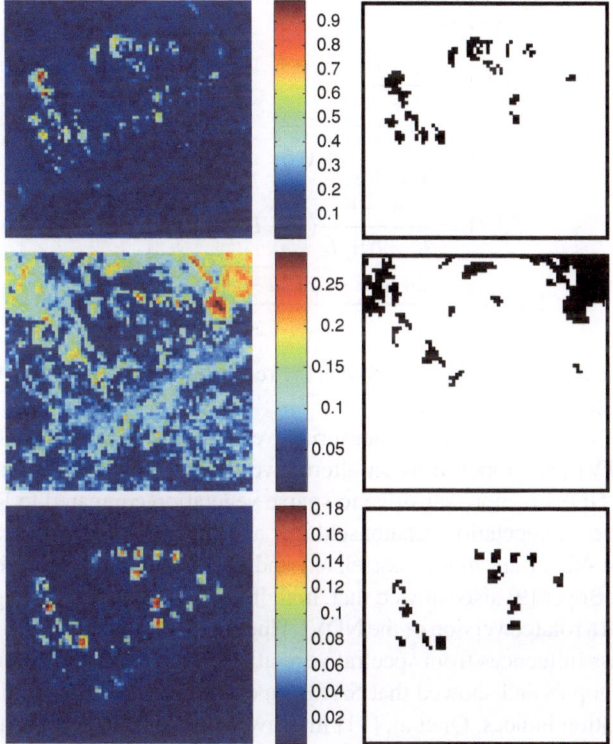

Fig. 3.2 Difference of KTT bands for the Adana image set. *First column* brightness, greenness, yellowness *second column* their thresholded versions

We provide the differences of brightness, greenness, and yellowness bands for the Adana image set in Fig. 3.2. As in the previous sections, we used Kapur's algorithm in finding the threshold value. As can be seen, the brightness and yellowness bands indicate changed regions.

3.3 Vegetation Index Differencing

Vegetation indices are obtained by transforming the multispectral data. They are used as a measure of vegetation which depends on the fact that vegetation absorbs most of the light in the red band and equally reflects the light in the infrared band. There are several vegetation indices in the literature, such as: Ratio vegetation index (RVI), normalized difference vegetation index (NDVI), transformed vegetation index (TVI), soil adjusted vegetation index (SAVI), and modified soil adjusted vegetation index (MSAVI). These are defined as

$$\text{RVI} = \frac{n}{r}, \tag{3.4}$$

$$\text{NDVI} = \frac{n-r}{n+r}, \tag{3.5}$$

$$\text{TVI} = \sqrt{\frac{n-r}{n+r} + 0.5}, \tag{3.6}$$

$$\text{SAVI} = \frac{n-r}{n-r+L}(1+L), \tag{3.7}$$

$$\text{MSAVI} = \frac{2n+1 - \sqrt{(2n+1)^2 - 8(n-r)}}{2}, \tag{3.8}$$

where n is the near infrared band and r is the red band. L in SAVI confirms the same bounds between NDVI and SAVI.

RVI is an earlier attempt for measuring vegetation density by band ratios [5]. NDVI and TVI are proposed as an alternative to RVI [6]. Jackson and Huete [7] showed that NDVI is more sensitive to sparse vegetation compared to RVI, but less sensitive to dense vegetation. Lautenschlager and Perry [8] mathematically showed that RVI and NDVI are highly correlated and thus contain the same information. Ünsalan and Boyer [9] also showed that, both indices can be taken as angles. Hence, RVI becomes a rotated version of the NDVI. Huete [10] introduced SAVI to minimize soil brightness influences from spectral vegetation indices. He studied on cotton and grassland canopies and showed that SAVI can eliminate variations originating from soil in vegetation indices. Qi et al. [11] improved SAVI and called the new index as MSAVI.

In terms of change detection, vegetation indices can be used for measuring the change of vegetation density in a given area with time. Any of the pixel-based methods described in this study can be applied using vegetation indices for an estimation of change in vegetation. Lunetta et al. [12] investigated the applicability of high resolution NDVI (250 m), MODUS NDVI, to land cover change detection. They studied on a geographic area where biological diversity and regrowth rates are high. Their results indicate that, up to 87 % correct change detection rates can be achieved. Guerra et al. [13] used MSAVI with bi-temporal Landsat TM images to identify vegetation changes. Guerra et al distinguished six types of land cover from a tropical area based on the spectral locations in the MSAVI space. They applied normalized image differencing for quantifying the change as

$$D = \frac{\text{MSAVI}_{t_2} - \text{MSAVI}_{t_1}}{\text{MSAVI}_{t_2} + \text{MSAVI}_{t_1}}. \tag{3.9}$$

In Fig. 3.3, we provide the difference images of RVI, NDVI, TVI, and SAVI for our Adana image set. Unfortunately, none of the indices provided useful results for change detection on the Adana image set.

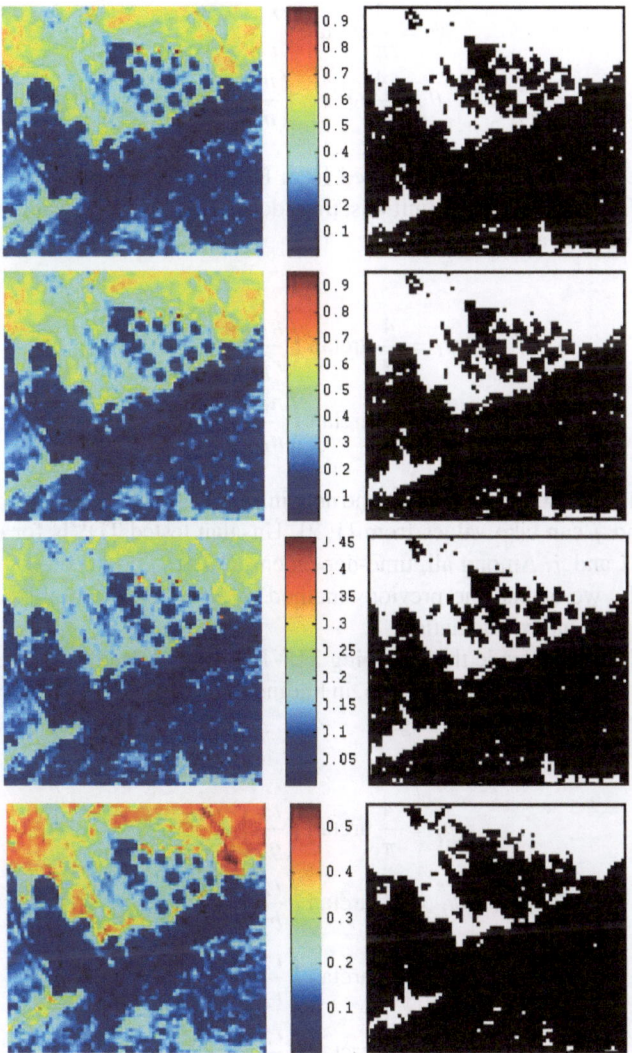

Fig. 3.3 Difference of RVI, NDVI, TVI, and SAVI for the Adana image set (*first column*). *Second column* their thresholded versions

3.4 Time-Dependent Vegetation Indices

A time-dependent vegetation index (TDVI) is a bi-temporal vegetation index calculated by using multispectral bands from t_1 and t_2 [14]. Red and near infrared bands from different times are involved in the same index formula. In a previous study, Ünsalan and Boyer [9] defined angle vegetation indices as

$$\psi = \frac{4}{\pi} \arctan\left(\frac{n}{r}\right), \tag{3.10}$$

$$\theta = \frac{4}{\pi} \arctan\left(\frac{n-r}{n+r}\right), \tag{3.11}$$

where ψ represents the angle obtained from RVI and θ represents angle obtained from NDVI. Based on these definitions, time-dependent form of the angle vegetation indices are defined as

$$\psi_t = \frac{4}{\pi} \arctan\left(\frac{n_i}{r_j}\right), \tag{3.12}$$

$$\theta_t = \frac{4}{\pi} \arctan\left(\frac{n_i - r_j}{n_i + r_j}\right), \tag{3.13}$$

where i and j are the time indices of the near infrared and red bands. For bi-temporal images i and j can take values from $\{1, 2\}$. Ünsalan tested TDVIs for every combination of i and j. Among all, time-dependent RVI (ψ_t) produced the best result. In this study, we extend the previous method by using all the multispectral band combinations in index calculation.

In Fig. 3.4, we provide the following TDVIs for our Adana image set. We used the percentile thresholding, by 97.5 %, in finding the threshold value. As can be seen, the results obtained are promising.

$$\psi_t' = \frac{4}{\pi} \arctan\left(\frac{r_2}{g_1}\right), \tag{3.14}$$

$$\psi_t'' = \frac{4}{\pi} \arctan\left(\frac{r_2}{b_1}\right), \tag{3.15}$$

$$\theta_t' = \frac{4}{\pi} \arctan\left(\frac{r_2 - g_1}{r_2 + g_1}\right), \tag{3.16}$$

$$\theta_t'' = \frac{4}{\pi} \arctan\left(\frac{r_2 - b_1}{r_2 + b_1}\right). \tag{3.17}$$

3.5 Color Invariants

Color invariants are transformations of the color images based on the correlations between multiple bands. These correlations are between the sensor response of the camera described in two parts: body reflection and specular reflection [15]. Gevers and Smeulders [16] evaluated the invariance of several transformations from RGB color space in terms of sensor responses based on the following criteria: viewing

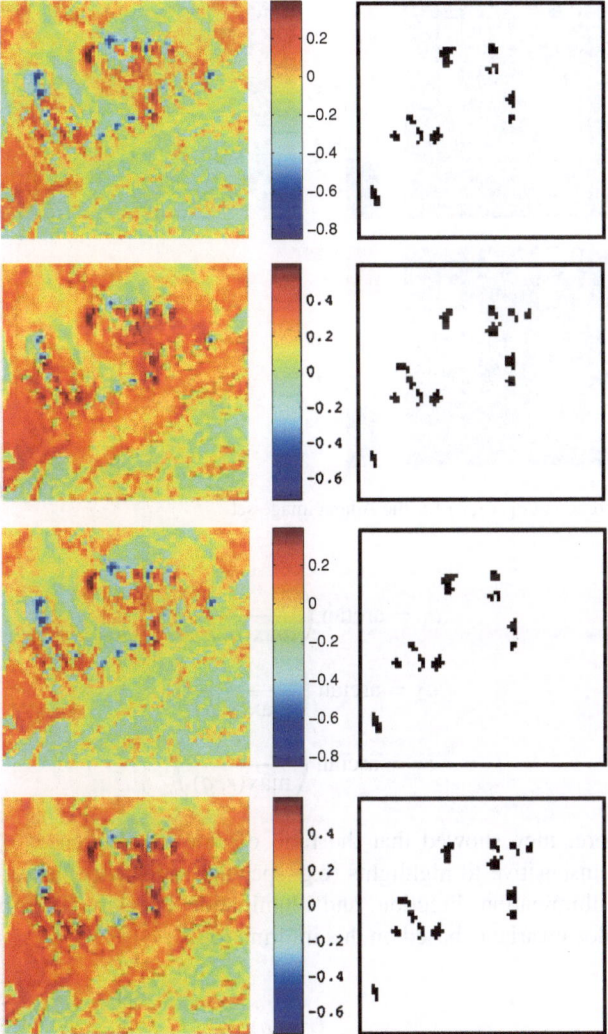

Fig. 3.4 ψ_t', ψ_t'', θ_t' and θ_t'' applied to the Adana image set (*first column*). *Second column* their thresholded versions

direction and object geometry, illumination direction, intensity of the illumination, and varying illumination color. They showed that the ratio of sum of the sensor responses are insensitive to surface orientation (object geometry), illumination direction, and illumination intensity. They proposed the following color invariants based on this information.

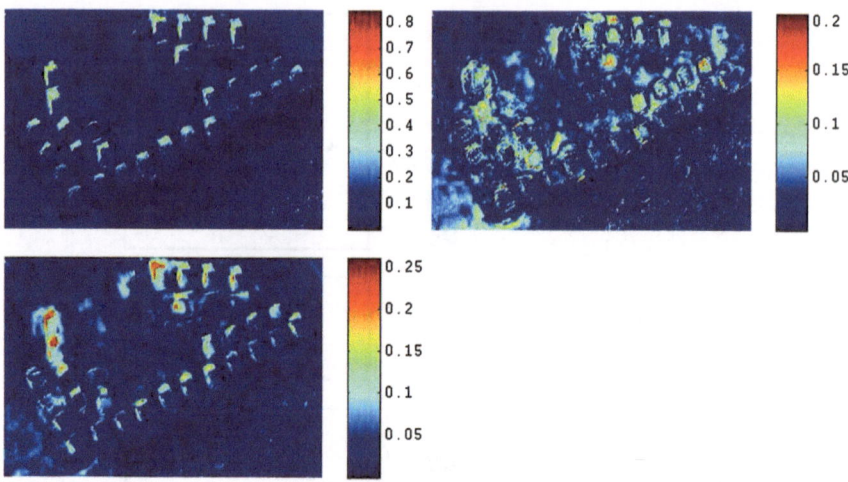

Fig. 3.5 Difference of c_1, c_2, c_3 for the Adana image set

$$c_1 = \arctan\left(\frac{r}{\max(g, b)}\right), \tag{3.18}$$

$$c_2 = \arctan\left(\frac{g}{\max(r, b)}\right), \tag{3.19}$$

$$c_3 = \arctan\left(\frac{b}{\max(r, g)}\right). \tag{3.20}$$

Furthermore, they showed that the ratio of sum of differences of the sensor responses is insensitive to highlights (e.g. specular reflectance) as well as surface orientation, illumination direction, and illumination intensity. They proposed the following color invariants based on this information.

$$l_1 = \frac{(r - g)^2}{(r - g)^2 + (r - b)^2 + (g - b)^2}, \tag{3.21}$$

$$l_2 = \frac{(r - b)^2}{(r - g)^2 + (r - b)^2 + (g - b)^2}, \tag{3.22}$$

$$l_3 = \frac{(g - b)^2}{(r - g)^2 + (r - b)^2 + (g - b)^2}. \tag{3.23}$$

Gevers and Smeulders pointed out the trade off between the discriminative power and the invariance of the color invariants. Suppose that color model A is invariant to illumination conditions w, x, y, and z, and color model B is invariant to illumination conditions y and z only. Under the illumination conditions where w, x, y, and z are uncontrolled (varies from sample to sample) color model A produces better

Fig. 3.6 Thresholded difference of c_2 for the Adana image set

results than B. On the other hand, the color model B gives better results than A under the illumination conditions where w and x are controlled and y and z are uncontrolled. Hence, while the invariance of the color model increases, its discriminative power decreases. For this reason, Gevers and Smeulders proposed color invariants for several invariance levels.

In change detection applications, highlights and illumination colors are controlled imaging conditions. Therefore, we need surface orientation and illumination intensity invariant models. Gevers and Smeulders recommended (c_1, c_2, c_3) color model for this type of invariance. As in the other transformation-based change detection methods, we can apply any of the pixel-based methods after transformation.

In Fig. 3.5, we provide the differences of c_1, c_2, c_3 color invariants for our Adana test image set. As can be seen, c_1 and c_3 color invariants mostly emphasize shadow regions in images. Therefore, they cannot be used for change detection directly. On the other hand, difference of the c_2 color invariant emphasizes significant changes such as missing and developing buildings. Hence, it can be used for change detection. In Fig. 3.6, we provide its thresholded version. As can be seen, while some of the important changes are kept, several other minor changes are enchanted. For this reason, the performance of c_2 difference is no better than simple image differencing.

References

1. Fung, T., LeDrew, E.: Application of principal components analysis to change detection. Photogram. Eng. Remote Sens. **53**(12), 1649–1658 (1987)
2. Kauth, R.J., Thomas, G.S.: The tasselled cap-a graphic description of the spectral-temporal development of agricultural crops as seen by Landsat. In: LARS Symposia, p. 159 (1976)
3. Seto, K.C., Woodcock, C.E., Song, C., Huang, X., Lu, J., Kaufmann, R.K.: Monitoring land-use change in the pearl river delta using landsat TM. Int. J. Remote Sens. **23**(10), 1985–2004 (2002)
4. Strang, G.: Linear Algebra and Its Applications. Academic Press, New York (1976)

5. Jordan, C.F.: Derivation of leaf-area index from quality of light on the forest floor. Ecology pp. 663–666 (1969)
6. Rouse, J.W., Haas, R.H., Schell, J.A.: Monitoring the Vernal Advancement and Retrogradation (Greenwave Effect) of Natural Vegetation. Texas A&M University, Texas (1974)
7. Jackson, R.D., Huete, A.R.: Interpreting vegetation indices. Prev. Vet. Med. **11**(3–4), 185–200 (1991)
8. Perry Jr, C.R., Lautenschlager, L.F.: Functional equivalence of spectral vegetation indices. Remote Sens. Environ. **14**(1–3), 169–182 (1984)
9. Ünsalan, C., Boyer, K.L.: Linearized vegetation indices based on a formal statistical framework. IEEE Trans. Geosci. Remote Sens. **42**(7), 1575–1585 (2004)
10. Huete, A.R.: A soil-adjusted vegetation index (Savi). Remote Sens. Environ. **25**(3), 295–309 (1988)
11. Qi, J., Chehbouni, A., Huete, A.R., Kerr, Y.H., Sorooshian, S.: A modified soil adjusted vegetation index. Remote Sens. Environ. **48**(2), 119–126 (1994)
12. Lunetta, R.S., Knight, J.F., Ediriwickrema, J., Lyon, J.G., Worthy, L.D.: Land-cover change detection using multi-temporal Modis Ndvi data. Remote Sens. Environ. **105**(2), 142–154 (2006)
13. Guerra, F., Puig, H., Chaume, R.: The forest-savanna dynamics from multi-date Landsat-TM data in Sierra Parima. Venezuela. Int. J. Remote Sens. **19**(11), 2061–2075 (1998)
14. Ünsalan, C.: Detecting changes in multispectral satellite images using time dependent angle vegetation indices. In: 3rd International Conference on Recent Advances in Space Technologies, pp. 345–348 (2007)
15. Shafer, S.A.: Using color to separate reflection components. Color Res. Appl. **10**(4), 210–218 (1985)
16. Gevers, T., Smeulders, A.: Color based object recognition. In: Image Analysis and Processing, pp. 319–326 (1997)

Chapter 4
Texture Analysis Based Change Detection Methods

Abstract In this chapter, we provide two texture based change detection methods. In both methods, we calculate the texture descriptors for bi-temporal images separately. In order to detect possible changes, we find their difference. We start with gray level co-occurrence matrix (GLCM) based texture descriptors next.

Keywords Texture analysis · Gray level co-occurrence matrix (GLCM) · Entropy

4.1 Gray Level Co-occurrence Matrix

Texture analysis focuses on the statistical explanation to the spatial distribution of the image pixels in a given image. There are several texture analysis methods proposed in the literature. A frequently used one is gray level co-occurrence matrix (GLCM) introduced by Haralick et al. [1]. GLCM entries are number of occurrences of spatial adjacency of gray tone values in an image. Adjacency is defined by the distance in pixel units.

The formal definition of GLCM is as follows. Let $I(x, y)$ be a grayscale image which takes values from the set $G = \{1, 2, \ldots, N_g\}$. The horizontal coordinate, x, of $I(x, y)$ takes values from the set $L_x = \{1, 2, \ldots, N_x\}$. The vertical coordinate, y, of $I(x, y)$ takes values from the set $L_y = \{1, 2, \ldots, N_y\}$. Then, the image $I(x, y)$ can be defined as a function from the set $L_x \times L_y$ to G.

The relative frequency of the spatial adjacency of gray tones i and j is mathematically defined for four directions as

$$
\begin{aligned}
P(i, j, d, 0°) = \#\{((k, l), (m, n)) \in (L_y \times L_x) \times (L_y \times L_x)| \\
k - m = 0, |l - n| = d, I(k, l) = i, I(m, n) = j\}
\end{aligned} \tag{4.1}
$$

M. İlsever and C. Ünsalan, *Two-Dimensional Change Detection Methods*,
SpringerBriefs in Computer Science, DOI: 10.1007/978-1-4471-4255-3_4,
© Cem Ünsalan 2012

$$P(i, j, d, 45°) = \#\{((k, l), (m, n)) \in (L_y \times L_x) \times (L_y \times L_x)|$$
$$(k - m = d, l - n = -d) \text{ or } (k - m = -d, l - n = d),$$
$$I(k, l) = i, I(m, n) = j\} \quad (4.2)$$

$$P(i, j, d, 90°) = \#\{((k, l), (m, n)) \in (L_y \times L_x) \times (L_y \times L_x)$$
$$||k - m| = 0, l - n = d, I(k, l) = i, I(m, n) = j\} \quad (4.3)$$

$$P(i, j, d, 135°) = \#\{((k, l), (m, n)) \in (L_y \times L_x) \times (L_y \times L_x)|$$
$$(k - m = d, l - n = d) \text{ or } (k - m = -d, l - n = -d),$$
$$I(k, l) = i, I(m, n) = j\}, \quad (4.4)$$

where d is the adjacency value, # sign indicates the number of occurrences under given conditions. As we mentioned earlier, GLCM entities are number of occurrences. Therefore, i and j are matrix indices and GLCM is an $N_g \times N_g$ matrix.

Haralick et al. calculated an average matrix from the four co-occurrence matrices as

$$P(i, j, d) = \frac{1}{4} \sum_{n=0}^{3} P(i, j, d, n \times 45°). \quad (4.5)$$

They used this matrix in calculating texture features. Thus, extracted features are rotation invariant for 45° of rotation.

Several texture features can be extracted from GLCM. The most useful ones for our purposes are

$$\text{Contrast: } \sum_i \sum_j (i - j)^2 P(i, j, d) \quad (4.6)$$

$$\text{Correlation: } \frac{\sum_i \sum_j (ij) P(i, j, d) - \mu_x \mu_y}{\sigma_x \sigma_y} \quad (4.7)$$

$$\text{Energy: } \sum_i \sum_j P^2(i, j, d) \quad (4.8)$$

$$\text{Inverse Difference Moment (IDM): } \sum_i \sum_j \frac{P(i, j, d)}{1 + (i - j)^2}, \quad (4.9)$$

where μ_x, μ_y, σ_x and σ_y are the mean and standard deviation of the marginal probability matrices $P_x = \sum_{j=1}^{N_g} P(i, j, d)$ and $P_y = \sum_{i=1}^{N_g} P(i, j, d)$.

As far as change detection is concerned, it is obvious that we can measure the amount of change by finding the difference in the amount of texture. Here, our approach must be slightly different from the pixel-based methods since in texture

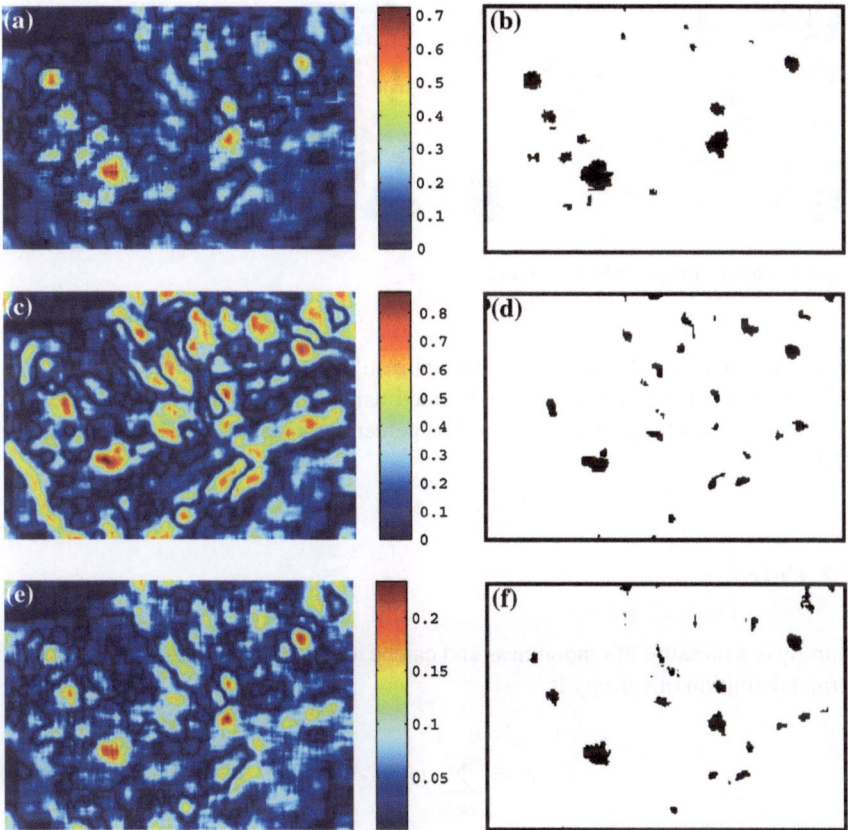

Fig. 4.1 Image differencing applied on GLCM features

analysis we study a region. One appropriate method is dividing the image into smaller windows and calculating texture features for these. This way, we can calculate the difference in texture by comparing the features on a window basis. As in pixel-based methods, images must be geometrically registered.

Tomowskia et al. [2] applied texture analysis methods to detect changes in satellite imagery. They used four change detection methods as image differencing, image rationing, image regression, and PCA. These methods are applied on texture features instead of the intensity values of the image. They benefit from four texture features as: contrast, correlation, energy, and IDM of GLCM. They calculated texture features in different window sizes ranging from 3×3 to 13×13. They reported the 13×13 windows to be the best. Tomowskia et al. tested their method on 16 change images. They reported the use of PCA with the energy feature produced best results.

We calculated contrast, energy, and inverse difference moment features for 13×13 windows on our Adana test image set. We applied image differencing, image rationing, and PCA using texture features. All change detection methods produced

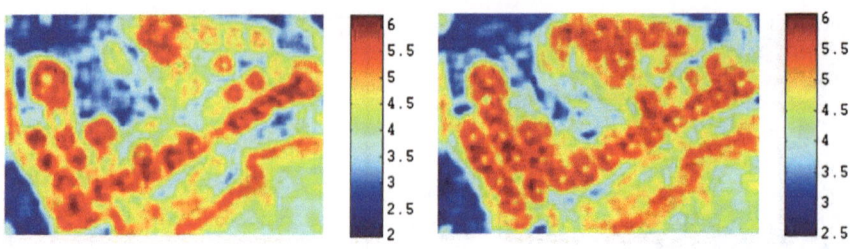

Fig. 4.2 Entropy images calculated for test images

similar results. Therefore, we only present results from image differencing in Fig. 4.1 where the difference images calculated using texture features are presented. In obtaining the change map, we applied percentile thresholding, percentile being 97.5 %.

4.2 Entropy

Entropy is a measure of randomness and can be used in quantifying texture [3]. The formal definition of entropy is

$$E = \sum_{i=0}^{N_g-1} p_i \log_2 p_i, \tag{4.10}$$

where p_i is the relative frequency of the intensity level i in the region and N_g is the number of possible intensity levels.

In using the entropy for change detection, we apply the following strategy. We calculate the entropy of our test images separately for 11×11 windows around each pixel. For the Adana test image set, we provide the obtained results in Fig. 4.2. As can be seen, developing regions are well emphasized by the second entropy image.

In our tests, we observed that taking the difference of entropy images does not produce expected results. Therefore, we first thresholded the images separately, then obtain their difference. We provide the thresholded (by 80 %) entropy images in Fig. 4.3a. Then, we obtain the difference by the binary XOR operation. To refine results, we removed noise in the difference image by applying morphological opening with a disk-shaped structuring element. We provide the final result in Fig. 4.3b. As can be seen, changed regions are detected by this method.

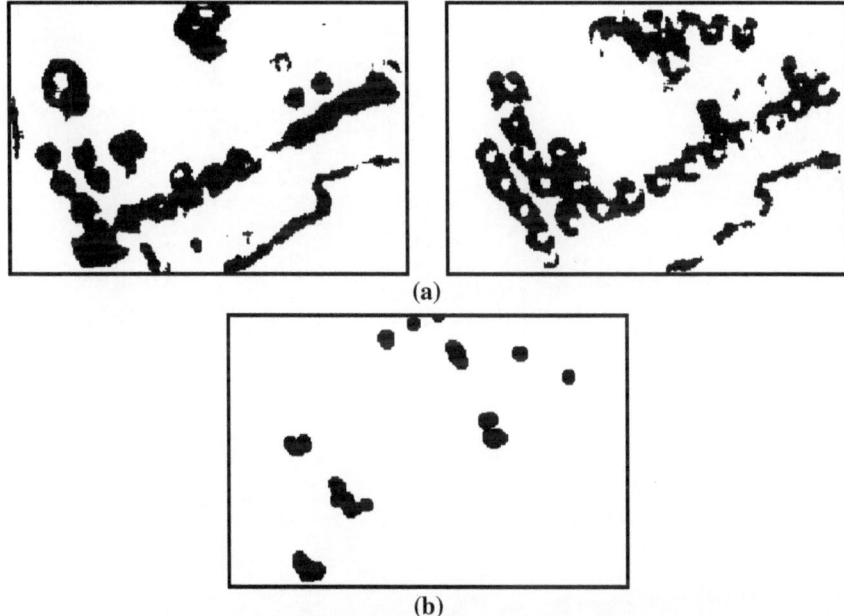

Fig. 4.3 Change detection by entropy texture feature

References

1. Haralick, R.M., Shanmugam, K., Dinstein, I.: Textural features for image classification. IEEE Trans. Syst. Man Cybern. **3**(6), 610–621 (1973)
2. Tomowski, D., Klonus, S., Ehlers, M., Michel, U., Reinartz, P.: Change visualization through a texture-based analysis approach for disaster applications. In: Proceedings of ISPRS Technical Commission VII Symposium-100 Years ISPRS Advancing Remote Sensing Science Sensing, vol. 38, pp. 263–268. Vienna, Austria (2010)
3. Gonzalez, R.C., Woods, R.E., Eddins, S.L.: Digital Image Processing Using MATLAB. Pearson Education, London (2004)

Chapter 5
Structure-Based Change Detection Methods

Abstract This chapter deals with change detection methods based on the structure information in bi-temporal images. We define the structure using six different methods. In the following sections, we explore each method in detail.

Keywords Structure · Edge detection · Gradient magnitude-based support regions (GMSR) · Matched filtering · Laplacian of Gaussian (LoG) · Mean shift · Segmentation · Solidity · Convex hull · Connected components · Local features · Scale invariant feature transform (SIFT) · Features from accelerated segment test (FAST) · Gaussian pdf · Graph matching · Shadow extraction

5.1 Edge Detection

The first method for structural change detection is based on edge information. We obtain edge pixels from two images using Canny's edge detection method [1]. For the registered image pairs, we expect to have correspondence between the edge maps of the two images. However, direct comparison is not feasible. Although the images can be registered, their looking angle may not be the same. Moreover, there may be shadow effects. Therefore, instead of finding the difference of both edge maps, we match the connected components between them [2]. We define a match between connected components from two images, if there is an overlap between them. Here, a partial overlap is also acceptable. Non-matched connected components in both images are taken as changed regions. In Fig. 5.1, we provide the edge maps obtained from the Adana image set. In Fig. 5.2, we provide the non-matched edge pixels. These represent the changed areas in the Adana image set.

M. İlsever and C. Ünsalan, *Two-Dimensional Change Detection Methods*,   41
SpringerBriefs in Computer Science, DOI: 10.1007/978-1-4471-4255-3_5,
© Cem Ünsalan 2012

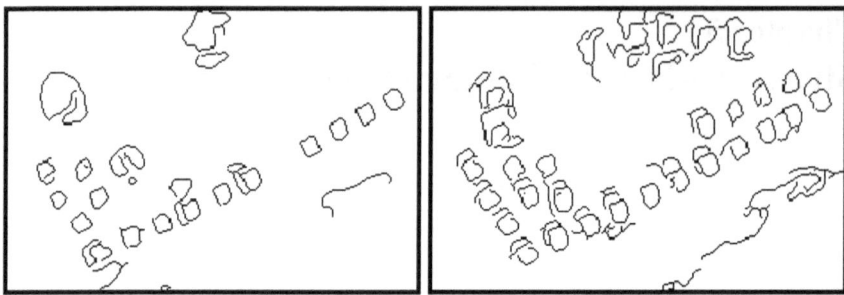

Fig. 5.1   Edge maps obtained from the Adana image set

Fig. 5.2   Change detection
results from the edge map
matching of the Adana image
set

5.2 Gradient-Magnitude-Based Support Regions

Similar to edge-based change detection, we can represent the edge information by
gradient-magnitude-based support regions. The Gradient-Magnitude-based Support
Regions (GMSR) is introduced in a previous study for land classification [3]. Here,
we apply the same methodology as we had done for edge detection with GMSR at
hand. We provide the GMSR obtained for both images in the Adana image set in
Fig. 5.3. We provide the change detection results in Fig. 5.4. As in the edge-based
method, changed regions are detected in this figure.

5.3 Matched Filtering

In the matched filtering approach, we assume a generic shape for buildings in the
image. Using the prototype for this shape, we detect buildings in images using
matched filtering (a standard method for digital communication systems) [4]. In this
study, we picked the Laplacian of Gaussian (LoG) filter as a generic building shape.
We apply it to both images and obtain high response regions, possibly representing
buildings. After thresholding, we apply the same methodology as we had done in the

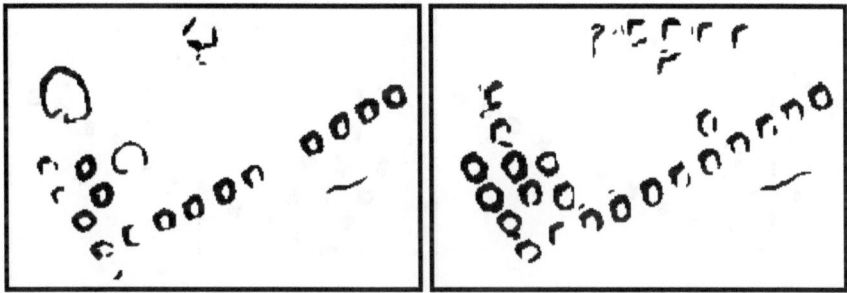

Fig. 5.3 GMSR obtained from the Adana image set

Fig. 5.4 Change detection results from the GMSR matching of the Adana image set

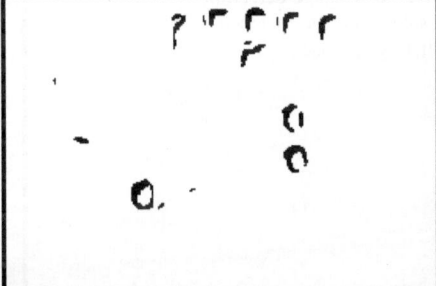

edge detection based approach. We provide the thresholded LoG responses for both Adana images in Fig. 5.5. We provide the change regions detected by this method in Fig. 5.6. Again, changed regions are detected fairly well.

5.4 Mean Shift Segmentation

Mean shift segmentation is introduced by Comanicu and Meer [5] as an application of feature space analysis. They referred to the density estimators for cluster analysis and in particular to the kernel density estimators. They showed that, mean shift vectors (obtained after the calculation of the density gradient) can be used for finding the local maxima points in the feature space. Their feature space formation consists of both spatial and spectral domain information. For segmentation, they first applied a mean shift-based edge preserving smoothing filter to the image. Then, they found the segments by delineating the feature space clusters which are the groups of mean shift filtered image pixels that belong to the same basin of attraction. A basin of attraction is defined as the set of all locations that converge to the same mode.

As far as change detection is concerned, similar to edge-based information, we consider segmentation of both images and detect changes based on segments. We refine segments based on their shape information using two region-based shape

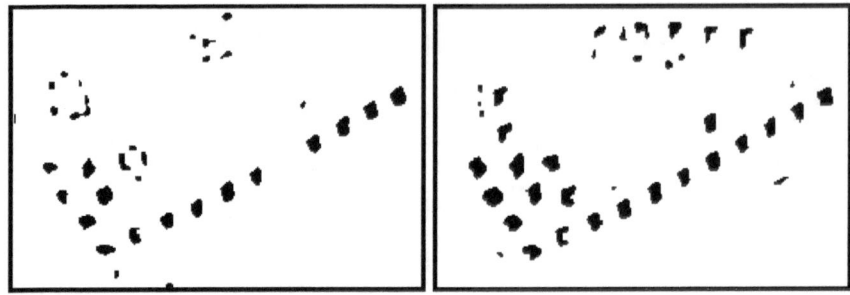

Fig. 5.5 Buildings detected from the Adana image set using matched filtering

Fig. 5.6 Change detection results from the matched filtering approach

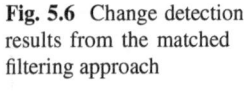

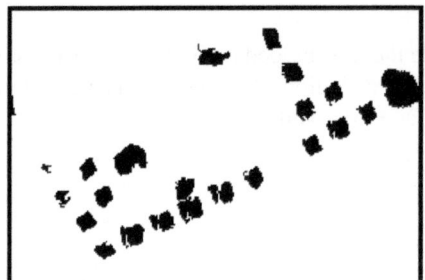

Fig. 5.7 Segments obtained from the Adana image set

descriptors as area and solidity [2]. Area is the number of pixels the region contains. Solidity is the ratio between the area and the area of the convex hull of the region. We eliminate segments having area greater than a threshold value and solidity less than a threshold value. Finally, we apply the methodology which we defined for the edge-based comparison to the segments to detect changes. We provide the segments obtained by mean shift clustering and shape refinement in Fig. 5.7. We provide the changed regions in Fig. 5.8. As in the previous structural methods, changed regions are detected fairly well by this approach.

Fig. 5.8 Change detection
results from the segments of
the Adana image set

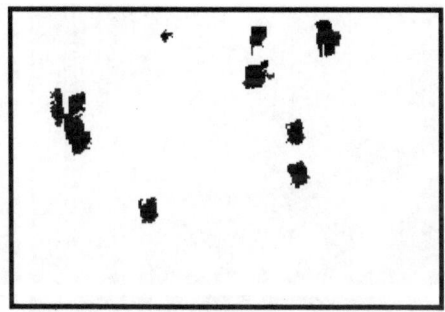

5.5 Local Features

We can describe any object in an image if we can find some distinctive features for it.
Once extracted, these features can be used for recognizing the object in another image
by comparing either their spatial locations or descriptors. Local feature detectors try
to isolate these features. Lowe [6] pointed out the distinctiveness of the features as
"The features must also be sufficiently distinctive to identify specific objects among
many alternatives. The difficulty of the object recognition problem is due in large
part to the lack of success in finding such image features. However, recent research
on the use of dense local features has shown that efficient recognition can often be
achieved by using local image descriptors sampled at a large number of repeatable
locations". He also refers to the repeatability of the local features in this quotation.
A feature detector should also be able to find distinctive features at the same location
under varying conditions such as scaling or illumination.

In this section, we propose a new change detection method based on local feature
matching. We use two widely used local feature detectors: Scale Invariant Feature
Transform (SIFT) and Features from Accelerated Segment Test (FAST) [6, 7]. SIFT
is a local feature detector with several valuable properties, such as invariance to
image scaling, translation, and rotation. It also has partial invariance to illumination
changes and affine or 3D projection. FAST is a high-speed feature detector.

Our change detection method has two steps: matching keypoints (local features)
and finding changed regions based on these. Keypoint matching is applied in a similar
manner for both SIFT and FAST. FAST keypoints are matched as follows. A keypoint
in the first image is assumed to be matched to a keypoint in the second image if they
are spatially close. Closeness norm is determined by fixed threshold values such as
dx and dy for two dimensions. In this way, if a keypoint in the first image is located at
(x, y), we search the existence of a keypoint in a region bounded by $[x - dx, x + dx]$
and $[y - dy, y + dy]$ in the second image. For SIFT keypoint matching, in addition to
the spatial constraint, we also compare the feature descriptor distances. Descriptor of
the keypoint in the first image is compared to the descriptor of the keypoints from the
previously defined bounded region. If there is a sufficiently close keypoint, then it is
assumed to be matched. No-matched keypoints are used for finding the change mask.

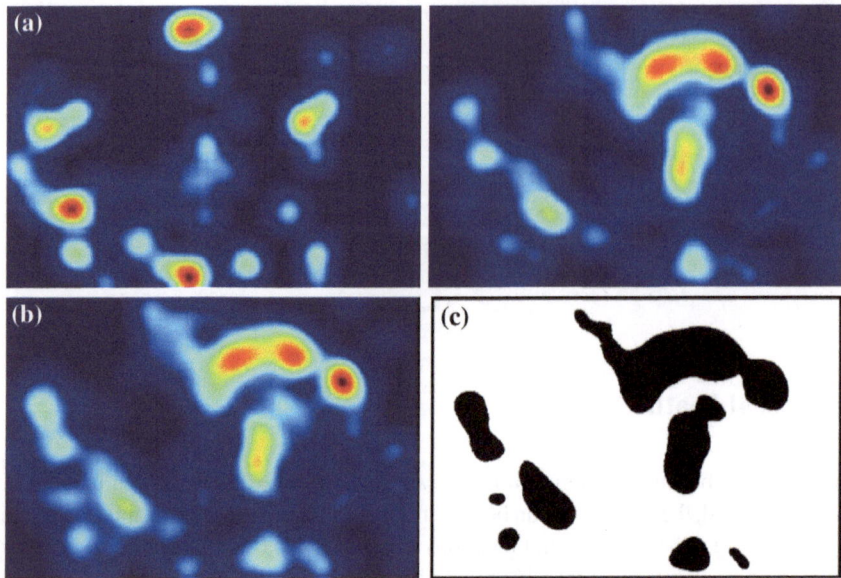

Fig. 5.9 Change detection using SIFT features. **a** Change density distributions for the Adana image set **b** Total change density distribution **c** Thresholded version

Centered at each non-matched keypoint, a Gaussian with a fixed variance is added onto the mask. Gaussians are added up to determine the change density distribution. Finally, a threshold is applied to this distribution to extract the change region.

We provide the change detection results for the Adana image set for three cases: matching only SIFT keypoints, matching only FAST keypoints and sum of separate results (SIFT and FAST). We provide the change detection results obtained by matching SIFT keypoints in Fig. 5.9. In Fig. 5.9a sum of the Gaussians centered at each non-matched keypoint is given. Separate change density distributions are added up resulting in total change as provided in Fig. 5.9b. The change density function is thresholded by 90th percentile, and the change regions are extracted. The result of this operation is provided in Fig. 5.9c. In a similar manner, we provide the change detection results obtained by matching FAST keypoints in Fig. 5.10. Finally, we provide the change detection results obtained after combining the SIFT and FAST results in Fig. 5.11. As can be seen, in both methods, changed regions are labeled correctly on the Adana image set.

5.6 Graph Matching

In this section, we propose a novel change detection algorithm based on the graph-based representation of the structural information. Our focus is detecting changes in a specific region using local features in a graph formalism. To represent the structure,

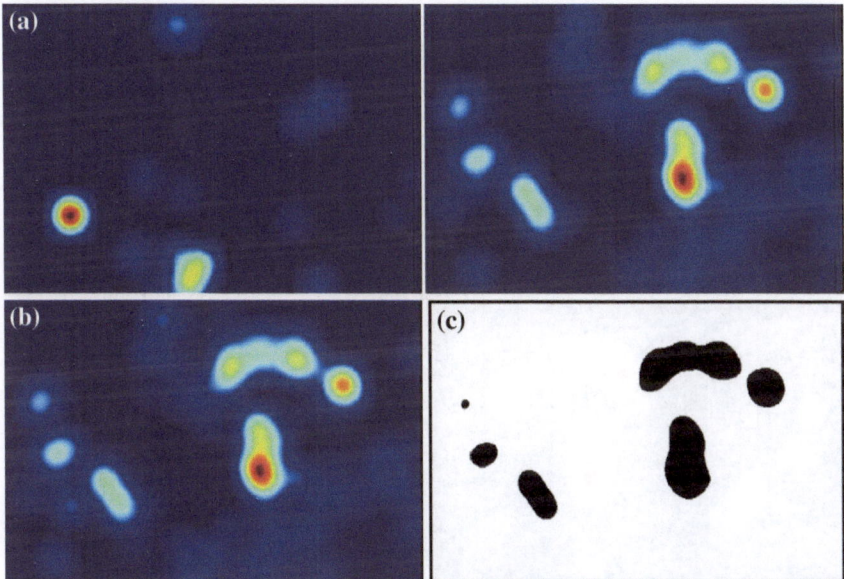

Fig. 5.10 Change detection using FAST features. **a** Change density distribution for the Adana image set. **b** Total change density distribution. **c** Thresholded version

we extract local features from both images using FAST. Then, we represent each local feature set (extracted from different images) in a graph form separately. This allows us to detect changes using a graph matching method.

To extract the structure information from local features, we represent them in a graph form. A graph G is represented as $G = (V, E)$, where V is the vertex set and E is the edge matrix showing the relations between these vertices. Here vertices are local features extracted by FAST. The edges are formed between them just by their distance. If a distance between two vertices are small, there will be an edge between them. In this study, we set this difference value to 10 pixels depending on the characteristics of the objects in the image.

As we form graphs from both images separately, we apply graph matching between them. In matching graphs, we apply constraints both in spatial domain and in neighborhood. We can summarize this method as follows. Let the graph formed from the first and second images be represented as $G_1(V_1, E_1)$ and $G_2(V_2, E_2)$. In these representations, $V_1 = \{f_1, \ldots, f_n\}$ holds the local features from the first image and $V_2 = \{s_1, \ldots, s_m\}$ holds the local features from the second image. We first take spatial constraints in graph matching. We assume that two vertices match if the spatial distance between them is smaller than a threshold. In other saying, f_i and g_j are said to be matched if $||f_i - g_j|| < \delta$ (δ being the threshold). This threshold adds a tolerance to possible image registration errors. Non-matched vertices from both graphs represent possible changed objects (represented by their local features). Since

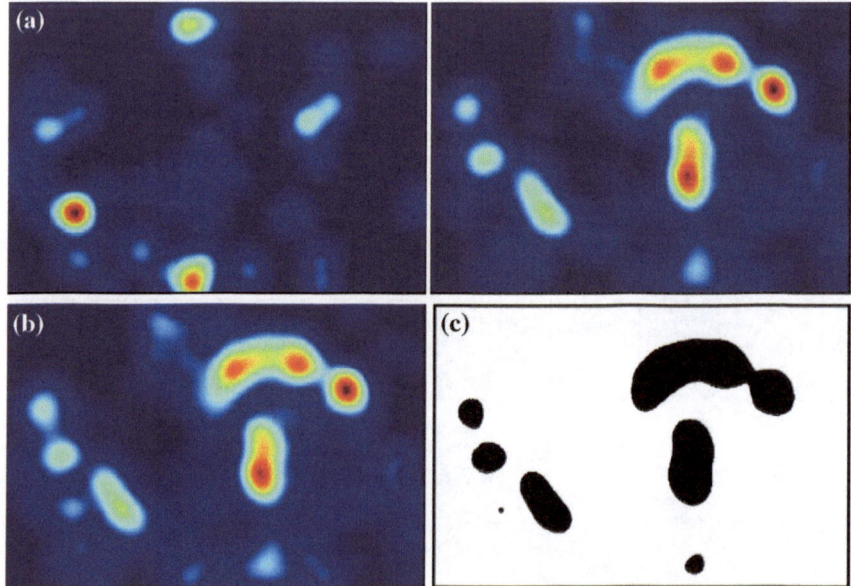

Fig. 5.11 Change detection using the combination of FAST and SIFT features. **a** Change density distribution for the Adana image set. **b** Total change density distribution. **c** Thresholded version

local features simply represent a single location, we add a circular tolerance to the non-matched ones to represent the changed area.

We can also add neighborhood information to graph matching. To do so, we first eliminate vertices having neighbors less than a number. Then, we match these refined vertices. This way, we eliminate some local features having no neighbors (possible noise regions).

We provide the local features extracted from the Adana image set in Fig. 5.12. Based on these, we provide the change detection results in Fig. 5.13. First, we did not apply any neighborhood constraint. Then, we refined vertices having less than three neighbors. Thus, we applied the neighborhood constraint. As can be seen, for both methods, changed regions are detected. However, as we add a neighborhood constraint, false alarms decreased dramatically.

5.7 Shadow Information

Shadow in a region gives indirect information about the elevation of the objects there. This information cannot be extracted by the methods that we considered so far in this study. Therefore, it is valuable. Elevation information can be helpful especially in urban monitoring, since man-made structures have height. In this study, we will use the shadow information for 2D change detection. We benefit from Sırmaçek and

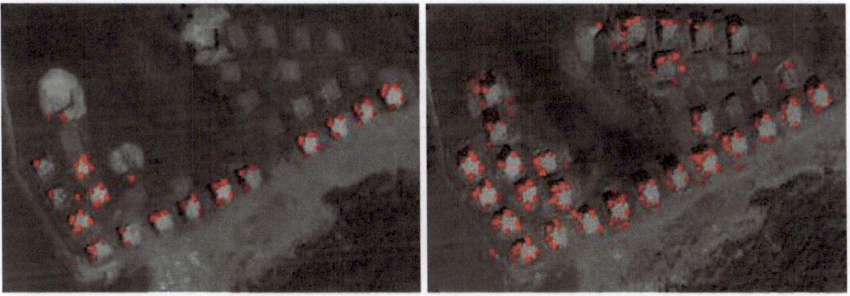

Fig. 5.12 Local features extracted by FAST which are used in graph formation

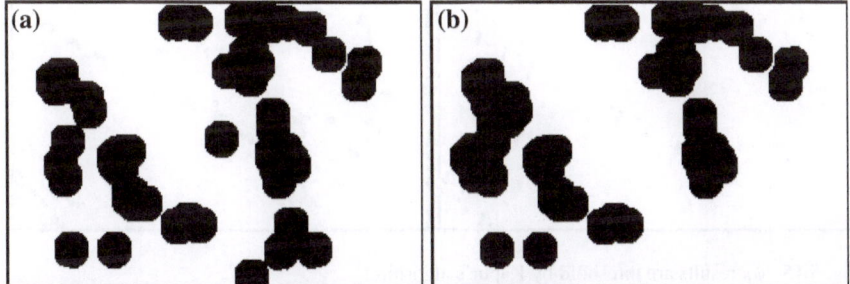

Fig. 5.13 Graph matching-based comparison results. **a** Using only spatial constraints **b** Adding three neighborhood constraint

Ünsalan's study [8] in extracting shadow regions. They used

$$\psi_b = \frac{4}{\pi} \arctan\left(\frac{b-g}{b+g}\right) \tag{5.1}$$

for shadow detection. In this equation, b and g are the blue and green bands of the image.

In detecting changes, we compare the shadow information extracted from times t_1 and t_2. Based on these, we estimate the spatial location of the change. In this method, images must be registered (as in most other change detection techniques) before comparison. Additionally, images must be taken at the same time of the year for a healthy comparison. As a final comment, the detected change locations are in terms of shadows. Therefore, they only represent the changed objects' shadows.

Our method has three steps. First, we extract shadow indicators by applying ψ_b to the test images. Then, we threshold the results and obtain the shadow regions. Finally, we detect shadow regions that do not exist in the other image. We provide the shadow indicators for our test images in Fig. 5.14. The thresholded (using Kapur's method) results are given in Fig. 5.15. Finally, we eliminate shadow regions that overlap with any shadow region in the other image. Remaining regions are accepted as shadow

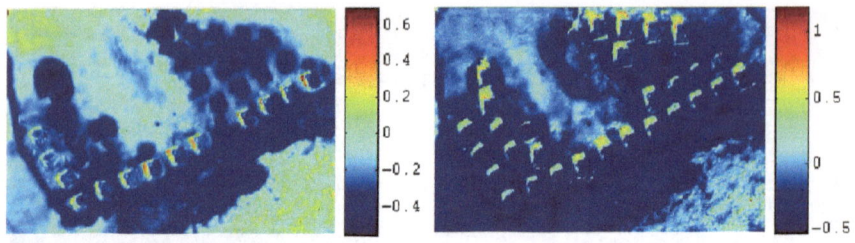

Fig. 5.14 ψ_b applied to test images

Fig. 5.15 ψ_b results are threshold by Kapur's algorithm

Fig. 5.16 Shadow differences

difference. We provide the obtained results in Fig. 5.16. While some shadow regions are missed, most of the regions are detected. We note that detected regions are located around the developing parts of the area.

References

1. Canny, J.: A computational approach to edge detection. IEEE Trans. Pattern Anal. Mach. Intell. **8**(6), 679–698 (1986)
2. Sonka, M., Hlavac, V., Boyle, R.: Image Processing, Analysis, and Machine Vision, vol. 2. PWS Publishing, Pacific Grove (1999)
3. Ünsalan, C.: Gradient-magnitude-based support regions in structural land use classification. IEEE Geosci. Remote Sens. Lett. **3**(4), 546–550 (2006)
4. Stark, H., Woods, J.W.: Probability and Random Processes with Applications to Signal Processing, vol. 3. Prentice Hall, Upper Saddle River (2002)
5. Comaniciu, D., Meer, P.: Mean shift: A robust approach toward feature space analysis. IEEE Trans. Pattern Anal. Mach. Intell. **24**(5), 603–619 (2002)
6. Lowe, D.G.: Object recognition from local scale-invariant features. In: The Seventh IEEE International Conference on Computer Vision, vol. 2, pp. 1150–1157 (1999)
7. Rosten, E., Drummond, T.: Machine learning for high-speed corner detection. ECCV **2006**, 430–443 (2006)
8. Sirmacek, B., Ünsalan, C.: Building detection from aerial images using invariant color features and shadow information. In: 23rd International Symposium on Computer and, Information Sciences, Istanbul, October 2008. pp. 1–5

Chapter 6
Fusion of Change Detection Methods

Abstract In previous chapters, we introduced several change detection methods working on different principles. They also had different assumptions. In this chapter, we provide a method to fuse the change maps obtained from these. We applied our fusion approach both in category (within a group) and inter-category (between groups) levels. We explore these methods next.

Keywords Decision fusion · Association

6.1 Fusion Methodology

Since our approach is combining the results (decisions) taken from different methods, it is a decision level fusion. Our fusion methodology consists of three operators: binary AND, binary OR, and binary association. Binary AND operator accepts that a pixel is truly classified as change only if all methods classified it as change. Binary OR operator accepts that a pixel is truly classified as change, if at least one method decided that way. In binary association, one method's decision is chosen as base. Decisions of other methods are compared with this decision on an object basis. Here, we find the objects (connected components) from one method which overlap the objects from the other method. In binary association, all decisions are compared with the base decision and found objects are added to the base decision. As a result, regions classified as change expand. The binary association result is this expansion plus the base decision.

We provide the fusion of the pixel-based methods with three operators along with the result from fuzzy XOR in Fig. 6.1. In calculating the binary association, we selected the fuzzy XOR change map as the base. We will observe the effect of fusion on the performance in the following chapter.

M. İlsever and C. Ünsalan, *Two-Dimensional Change Detection Methods*,
SpringerBriefs in Computer Science, DOI: 10.1007/978-1-4471-4255-3_6,
© Cem Ünsalan 2012

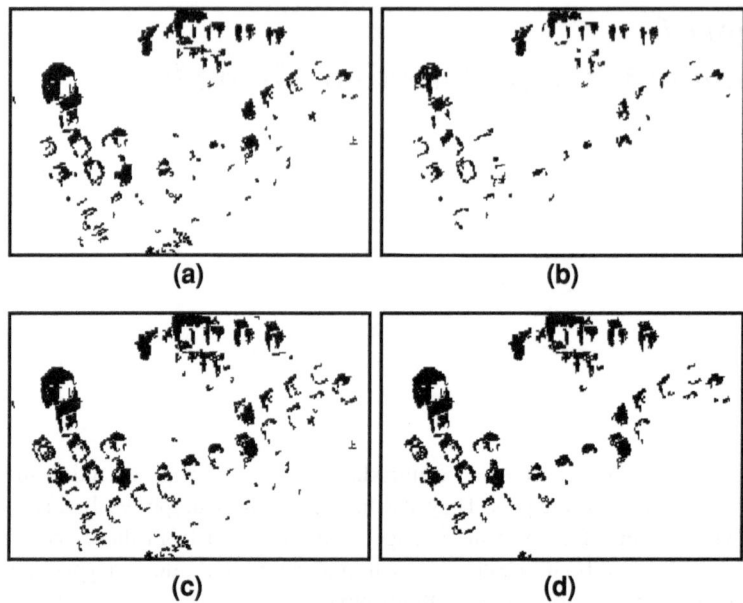

(a)   (b)

(c)   (d)

Fig. 6.1 Fusion of pixel-based change detection results. **a** Fuzzy XOR change map **b** Fusion by the AND opreator **c** Fusion by the OR operator **d** Fusion by association

6.2 Category Level Fusion

In category level fusion, all methods from one change detection method category (e.g. pixel-based methods) are combined. We provide the association of the structure-based results for the Adana test image set in Fig. 6.2. In this fusion example, decision of the graph matching method is taken as the base decision.

6.3 Inter-Category Level Fusion

In inter-category level fusion, for each method category, we use the results from the fusion operator with the best performance (to be provided in Sect. 7.2.6). We restricted inter-category fusion only to association, since fusion of the structure-based methods category produces results only in object basis. In Fig. 6.3, we provide the fusion of pixel and structure-based change detection results.

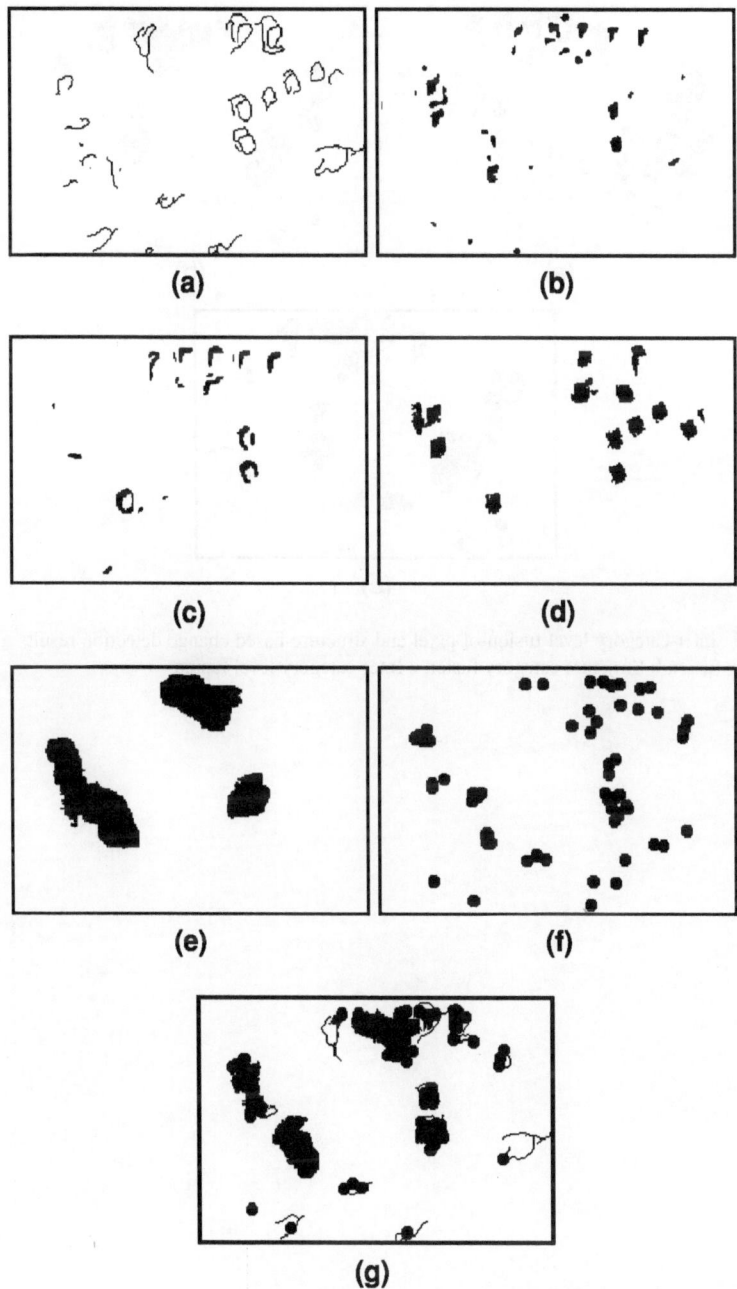

Fig. 6.2 Category level fusion of structure-based change detection results **a** Edge detection **b** Matched filtering **c** GMSR **d** Segmentation **e** Local features **f** Graph matching **g** Fusion by association

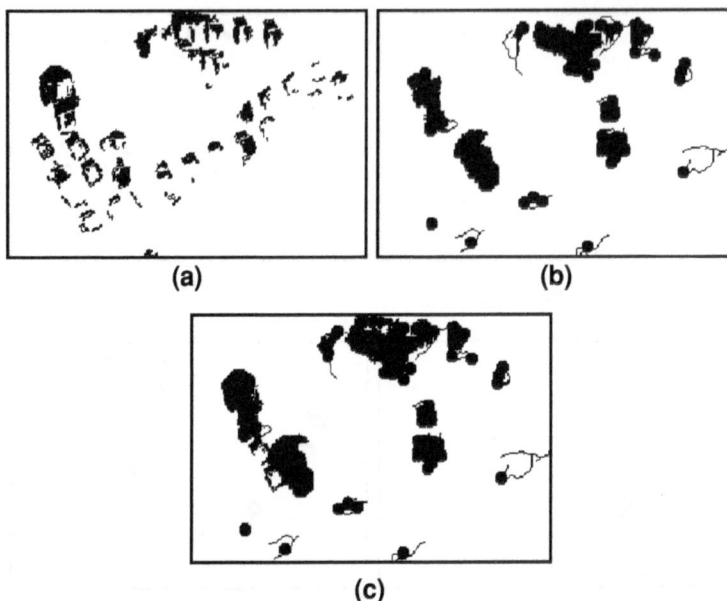

(a)   (b)

(c)

Fig. 6.3 Inter-category level fusion of pixel and structure-based change detection results **a** Pixel category fusion **b** Structure category fusion **c** Inter-category level fusion

Chapter 7
Experiments

Abstract In this chapter, we provide experimental results obtained from all change detection methods considered in this study. We first explain the data set used in the experiments. Then, we provide performance results in tabular form for each change detection method in detail.

7.1 The Data Set

Our data set consists of images extracted from high-resolution Ikonos images. These are acquired from the particular regions of Ankara and Adana in four different times between years 2001 and 2004. Besides using three band panchromatic images, we also used their four band multispectral versions. Panchromatic Ikonos images have one meter resolution. Corresponding multispectral images have four meter resolution and contain red, green, blue, and near infrared spectral bands. In our data set, 18 image pairs are panchromatic (labeled with letter P in suffix) and 17 image pairs are multispectral (labeled with letter M in suffix). Our test images are geometrically registered. They are also radiometrically normalized as discussed in Sect. 2.1. We provide a sample set of our test images in Figs. 2.2 and 7.1.

7.2 Performance Tests

In evaluating each method, we have a manually generated ground truth image set. In forming the ground truth, we specifically focused on urban area changes. Therefore, we mainly labeled the changed regions which occurred in the man-made structures such as buildings and road segments. In Fig. 7.2, we provide three ground truth images generated for $AdanaP_2$, $AnkaraP_3$, and $AnkaraP_{10}$ image pairs as an example.

M. İlsever and C. Ünsalan, *Two-Dimensional Change Detection Methods*,
SpringerBriefs in Computer Science, DOI: 10.1007/978-1-4471-4255-3_7,
© Cem Ünsalan 2012

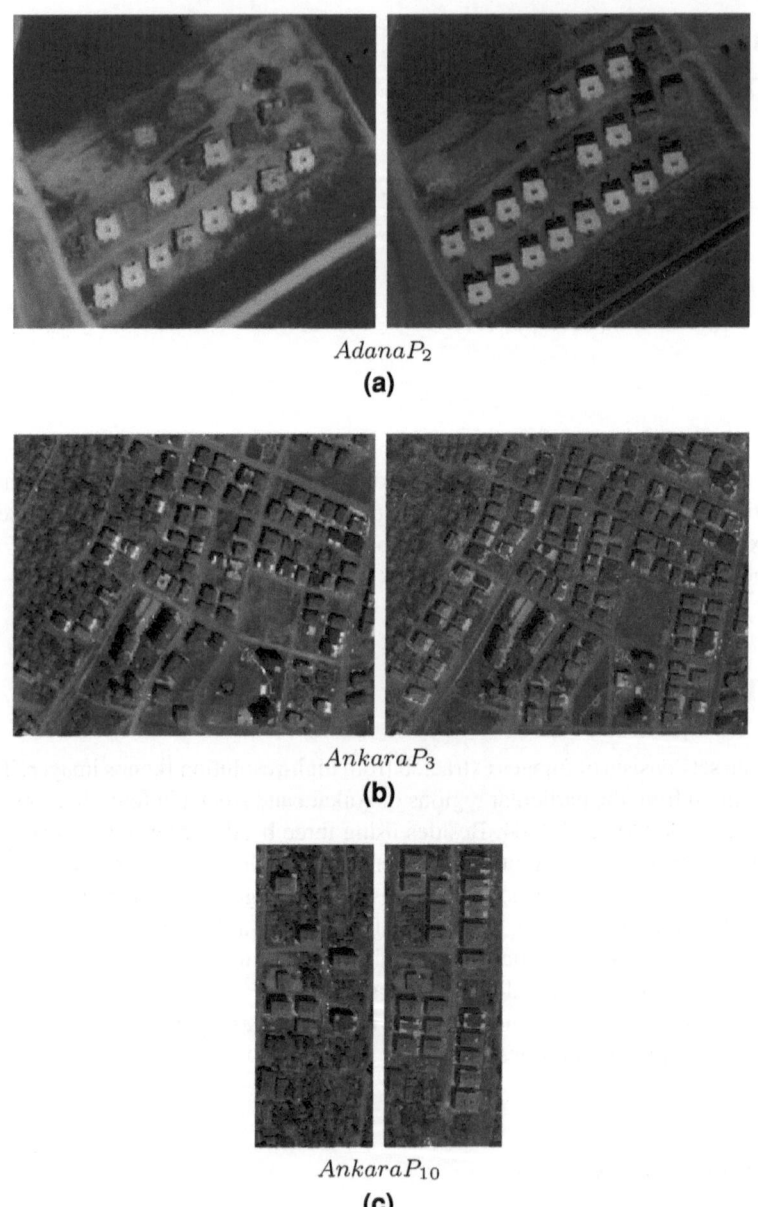

AdanaP₂
(a)

AnkaraP₃
(b)

AnkaraP₁₀
(c)

Fig. 7.1 Three sample image pairs acquired from Adana and Ankara **a** *Adana P₂* **b** *Ankara P₃* **c** *Ankara P₁₀*

In Tables 7.1 and 7.2, we provide number of ground truth pixels (GT) and number of image pixels (IP) for each image pair. 'GT Pixels/Image Pixels' ratio (GT/IP)

Fig. 7.2 Ground truth images for three test image pairs for *Adana* P_2, *Ankara* P_3, and *Ankara* P_{10}

Table 7.1   Ground truth pixel information for the panchromatic test image set

	Image size	GT	IP	GT/IP
Adana P_1	270 × 180	5893	48600	0.1213
Adana P_2	260 × 220	5500	57200	0.0962
Adana P_3	200 × 240	2087	48000	0.0435
Adana P_4	200 × 200	7180	40000	0.1795
Adana P_5	260 × 400	8383	104000	0.0806
Adana P_6	250 × 280	11185	70000	0.1598
Adana P_7	350 × 350	18296	122500	0.1494
Adana P_8	290 × 440	9549	127600	0.0748
Ankara P_1	700 × 700	97684	490000	0.1994
Ankara P_2	468 × 477	41573	223236	0.1862
Ankara P_3	550 × 445	40931	244750	0.1672
Ankara P_4	780 × 350	36525	273000	0.1338
Ankara P_5	850 × 760	187970	646000	0.2910
Ankara P_6	490 × 520	89876	254800	0.3527
Ankara P_7	340 × 330	36074	112200	0.3215
Ankara P_8	250 × 310	15800	77500	0.2039
Ankara P_9	360 × 340	18878	122400	0.1542
Ankara P_{10}	160 × 400	21215	64000	0.3315

indicates the amount of change involved in the specific image pair. We provide the total panchromatic and multispectral image pixels in Table 7.3.

Our comparisons are based on the intersections, unions, and complements of the ground truth and the result set. We use four measures as True Positive (TP), False Positive (FP), True Negative (TN), and False Negative (FN). In terms of ground truth set (GT) and the result set (RE) these quantities are defined as

Table 7.2 Ground truth pixel information for the multispectral test image set

	Image size	GT	IP	GT/IP
$Adana M_1$	65×60	426	3900	0.1092
$Adana M_2$	120×140	1386	16800	0.0825
$Adana M_3$	140×260	13993	36400	0.3844
$Adana M_4$	75×70	381	5250	0.0726
$Adana M_5$	100×90	2063	9000	0.2292
$Adana M_6$	90×90	1019	8100	0.1258
$Adana M_7$	120×100	1571	12000	0.1309
$Adana M_8$	90×70	316	6300	0.2089
$Ankara M_1$	290×210	4724	60900	0.0776
$Ankara M_2$	150×250	6454	37500	0.1721
$Ankara M_3$	80×155	4276	12400	0.3448
$Ankara M_4$	80×70	1296	5600	0.2314
$Ankara M_5$	90×75	2279	6750	0.3376
$Ankara M_6$	65×100	1610	6500	0.2477
$Ankara M_7$	100×80	1084	8000	0.1355
$Ankara M_8$	90×90	1351	8100	0.1668
$Ankara M_9$	100×100	1268	10000	0.1268

Table 7.3 Ground truth total pixel information for panchromatic images and multispectral images

Image type	Total GT	Total IP	Total GT/IP
Panchromatic	654599	3125786	0.2094
Multispectral	46497	253500	0.1834

$$TP = GT \cap RE \qquad\qquad (7.1)$$

$$FP = RE - GT \qquad\qquad (7.2)$$

$$TN = (GT \cup RT)^c \qquad\qquad (7.3)$$

$$FN = GT - RE \qquad\qquad (7.4)$$

where '$-$' sign is the set-theoretic difference and superscript 'c' is the set complement. In Fig. 7.3, we demonstrate these four set regions.

In terms of pixel-based comparison, TP represents the number of change pixels correctly detected. FP represents the number of no-change pixels incorrectly labeled as change by the method. TN represents the number of no-change pixels correctly detected. FN represents the number of change pixels incorrectly labeled as no-change by the method. Based on these quantities, we benefit from three performance criteria as: the percentage correct classification (PCC), the Jaccard coefficient (Jaccard), and the Yule coefficient (Yule) [1]. These are defined as

$$PCC = \frac{TP + TN}{TP + FP + TN + FN} \qquad\qquad (7.5)$$

$$Jaccard = \frac{TP}{TP + FP + FN} \qquad\qquad (7.6)$$

Fig. 7.3 Pixel correspondence between the ground truth and the result set

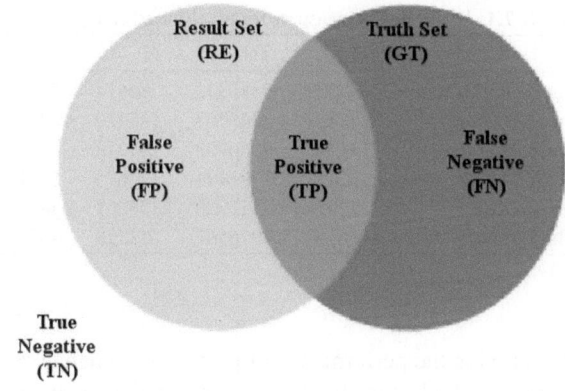

$$\text{Yule} = \frac{\text{TP}}{\text{TP} + \text{FP}} + \frac{\text{TN}}{\text{TN} + \text{FP}} - 1 \qquad (7.7)$$

PCC is the most common performance measure. Literally, it is the comparison of the truly found pixels to the whole pixel set. However, this measure is not sufficient for our comparisons for images containing little change. From Tables 7.1 and 7.2, we can see that 'GT/IP' value can drop as low as 0.043 which indicates minor change. In such cases, any method can get very high PCC value by just labeling all pixels as negative. TN dominates the PCC measure and it can easily reach the value one. To avoid this, we use the Jaccard measure. This measure excludes TN in its formulation. PCC and Jaccard measures are in the range [0, 1]. The Yule measure is in the range [−1, 1]. The higher these measures, the better the performances.

7.2.1 Pixel-Based Change Detection Methods

We start providing our test results with pixel-based methods. They use only panchromatic image pairs. Therefore, in these tests we used only 18 image pairs from our database. Instead of giving results for each image separately, we provide the total image statistics. To this end, we sum TN, TP, FP, and FN for each image and then calculate the performance measures. One exception to this is the background subtraction method. It requires more than two images for background formation. In our study, for most of our test set, we have four images from the same region. Unfortunately, for five test image sets we do not have images from four different times. Therefore, we used only 13 of the 18 regions in background formation. Furthermore, instead of giving four subtraction results separately for background formation based method, we only provide sum of the first and fourth subtractions. This is because, at the two extremes of our time interval (first and fourth images) represent most of the change.

Table 7.4 Performance measures for pixel-based change detection methods

	TP	TN	FP	FN	PCC	Jaccard	Yule
Image differencing	139599	2386573	84614	515000	0.8082	0.1888	0.4451
Image ratioing	171727	2345767	125420	482872	0.8054	0.2202	0.4072
Image regression	141249	2366909	104278	513350	0.8024	0.1861	0.3971
CVA	132453	2393494	77693	522146	0.8081	0.1809	0.4512
Fuzzy XOR	72034	2435920	35267	582565	0.8023	0.1044	0.4783
Background sub.	148360	1976467	92328	457331	0.7945	0.2125	0.4285

We provide the performance of pixel-based change detection methods in Table 7.4. For all methods, threshold values are obtained using Otsu's algorithm.

As can be seen in Table 7.4, image differencing, image regression, and CVA methods produced similar results. Image ratioing produced better Jaccard and worse Yule values. This indicates that, while it finds more TP pixels compare to other methods, it labels noise pixels as well. Fuzzy XOR method produced a similar PCC value compared to other methods, but it failed to find comparable TP pixels using Otsu's method.

7.2.2 Transformation Based Change Detection Methods

We provide performance test results for transformation based methods in Table 7.5. Again, Otsu's algorithm is used in determining the threshold values. First four rows of the table are principal component results. Since we used four band multispectral images in PCA, we found four principal components. We used separate rotation-type PCA (Sect. 3.1).

As can be seen in Table 7.5, the first PC produced results close to pixel-based methods. Other principal components produced poor results, even the second PC produced a negative Yule value. Low PCC and Yule values are due to the noise generated by these principal components. For the KTT method, the brightness band performed best. Other band results are comparable, but not close to pixel-based methods in terms of the Jaccard measure. Vegetation indices RVI, NDVI, TVI, and SAVI performed fairly well. We tested four time-dependent vegetation indices (TDVI). ψ_t'' and θ_t'' which use red and blue bands produced better results. For these, high PCC and Yule values indicate that threshold values can be decreased for higher Jaccard value. Finally, the color invariant based change detection method produced poor results.

7.2.3 Texture-Based Change Detection Methods

We provide the texture-based change detection results in Table 7.6. Here, we also use Otsu's algorithm in threshold selection. For entropy method, an 11×11 sliding window is used. For GLCM, a 13×13 sliding window is used. As for GLCM fea-

Table 7.5 Performance measures for transformation based change detection methods

	TP	TN	FP	FN	PCC	Jaccard	Yule
First PC	8859	199184	7819	37638	0.8207	0.1631	0.3722
Second PC	3519	190445	16558	42978	0.7651	0.0558	−0.0088
Third PC	3928	197725	9278	42569	0.7955	0.0704	0.1203
Fourth PC	2409	197068	9935	44088	0.7869	0.0427	0.0123
KTT-brightness	9750	201605	5398	36747	0.8337	0.1879	0.4895
KTT-greenness	8122	198718	8285	38375	0.8159	0.1483	0.3332
KTT-yellowness	8022	201363	5640	38475	0.8260	0.1539	0.4268
RVI	9409	199450	7553	37088	0.8239	0.1741	0.3979
NDVI	9387	199579	7424	37110	0.8243	0.1741	0.4016
TVI	9830	199268	7735	36667	0.8248	0.1813	0.4042
SAVI	8646	199113	7890	37851	0.8196	0.1590	0.3631
ψ_t'	4173	205078	1925	42324	0.8254	0.0862	0.5132
ψ_t''	6093	205327	1676	40404	0.8340	0.1265	0.6198
θ_t'	4173	205078	1925	42324	0.8254	0.0862	0.5132
θ_t''	6093	205327	1676	40404	0.8340	0.1265	0.6198
c_2	12282	2440180	31007	642317	0.7846	0.0179	0.0753

Table 7.6 Performance measures for texture-based change detection methods

	TP	TN	FP	FN	PCC	Jaccard	Yule
Entropy	111029	2344253	126934	543570	0.7855	0.1421	0.2784
Contrast	179817	2190678	280509	474782	0.7584	0.1923	0.2125
Energy	244870	1978734	492453	409729	0.7114	0.2135	0.1606
IDM	276922	1933660	537527	377677	0.7072	0.2323	0.1766

tures, Contrast, Energy, and Inverse Difference Moment (IDM) are calculated. PCC values are lower than all pixel methods for all texture-based methods. Yule values are also poor compared to pixel-based methods. In terms of the Jaccard measure, IDM produced the best result but with a very poor PCC and Yule value.

7.2.4 Comparison of Thresholding Algorithms

Up to this point, we only provided change detection results using Otsu's thresholding algorithm. Now, we evaluate the effect of Kapur's algorithm and percentile thresholding in change detection performance. Kapur's algorithm determines the threshold value automatically. However, a percent value should be determined for percentile thresholding. Therefore, we tested change detection methods starting from 99.0 to 75.0 %. Decreasing the percentile value added more TP to the results. This led to an increase in the *Jaccard* measure. Rate of increase was high for the higher percentile values, and it gets lower when we drop the percentile to 75 %. We demonstrate this behavior for image differencing method in Table 7.7. In this table, we focused

Table 7.7 Effect of the percentile value on the performance measures

Percentile (%)	TP	TN	FP	FN	PCC	Jaccard	Yule
99.0	17358	2465098	6089	637241	0.7942	0.0263	0.5349
97.5	43815	2451981	19206	610784	0.7985	0.0650	0.4958
95.0	86351	2425828	45359	568248	0.8037	0.1234	0.4658
90.0	164595	2360764	110423	490004	0.8079	0.2152	0.4266
85.0	231113	2279769	191418	423486	0.8033	0.2732	0.3903
80.0	286262	2184580	286607	368337	0.7905	0.3041	0.3554
75.0	365591	1954999	473481	283115	0.7541	0.3258	0.3092

on the *Jaccard* measure. As a reasonable level, we pick the percentile to be 85.0 %
and applied it to all our methods.

We compared percentile thresholding, Otsu's method, and Kapur's method using
results from pixel-based methods in Fig. 7.4. We compared the PCC, Jaccard, and
Yule values produced by thresholding algorithms. If a thresholding algorithm produced better PCC, Jaccard, and Yule value compared to another algorithm, then we
say that it performed better for the specific change detection method. Otsu's algorithm
performed better than Kapur's when used for image differencing, image regression,
and CVA. In this table, we also take the specific thresholding method for the fuzzy
XOR-based change detection method. We label it as *std* in the chart. As can be seen,
for this specific method, thresholds should be selected by the standard deviation as
mentioned in Sect. 2.6.

We also compared thresholding algorithms via PCA and KTT methods in Fig. 7.5.
Here, Otsu's algorithm performed better than Kapur's when used for KTT brightness
and greenness bands.

We next compared thresholding algorithms using the results from vegetation
indices, TDVIs, and color invariant in Fig. 7.6. For RVI and NDVI, Kapur's algorithm performed better than others. Otsu's algorithm performed better than Kapur's
when used for TVI and SAVI. Percentile thresholding performed better than Kapur's
algorithm for TDVI and the color invariant.

We finally compared thresholding algorithms in terms of texture analysis based
methods in Fig. 7.7. Otsu's algorithm performed better than Kapur's when used for
entropy based change detection. Percentile thresholding performed better than Otsu's
algorithm when used for contrast feature of the GLCM.

Based on all the above tests, we decided to use Otsu's thresholding algorithm for
our change detection methods. However, for specific applications, other thresholding
methods may also be used. Therefore, there is no clear winner among thresholding
algorithms.

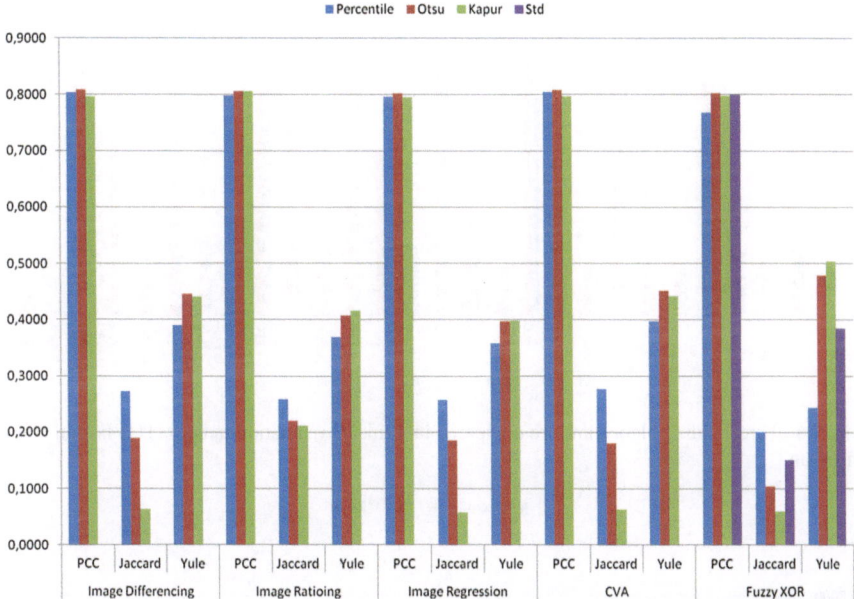

Fig. 7.4 Thresholding algorithms are compared in terms of pixel-based methods

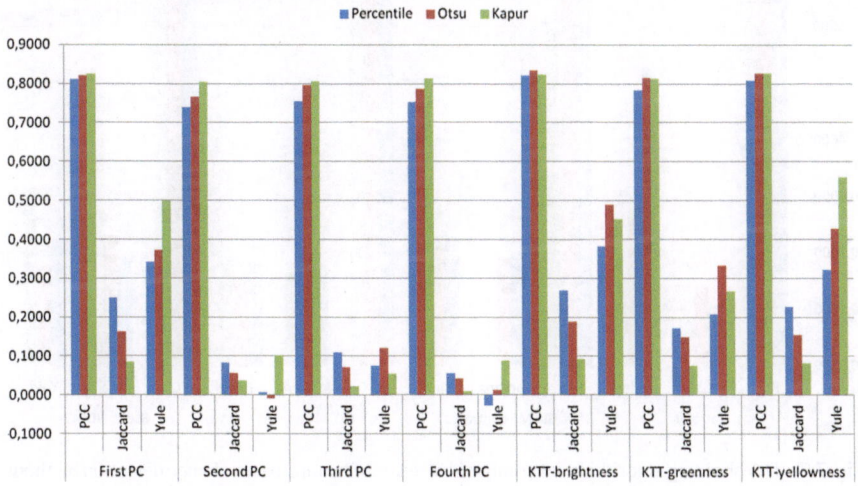

Fig. 7.5 Thresholding algorithms are compared in terms of PCA and KTT

7.2.5 *Structure-Based Change Detection Methods*

So far, we provided results based on pixel-based comparison. However, this is not
the case for structural methods, since each generate a different type of output.

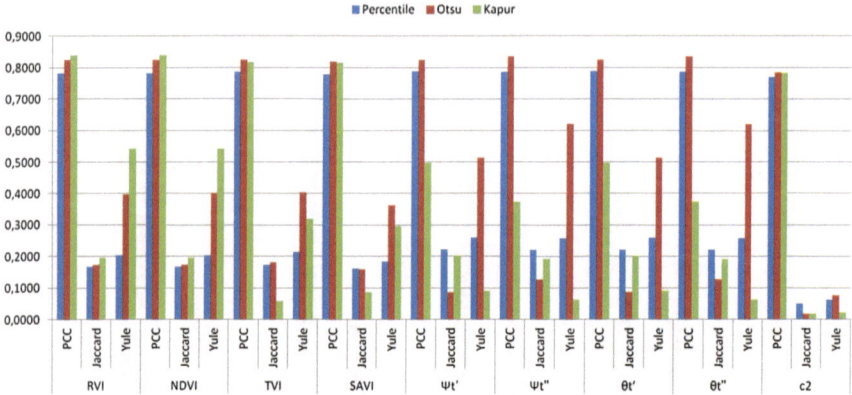

Fig. 7.6 Thresholding algorithms are compared in terms of vegetation indices, TDVIs, and color invariant

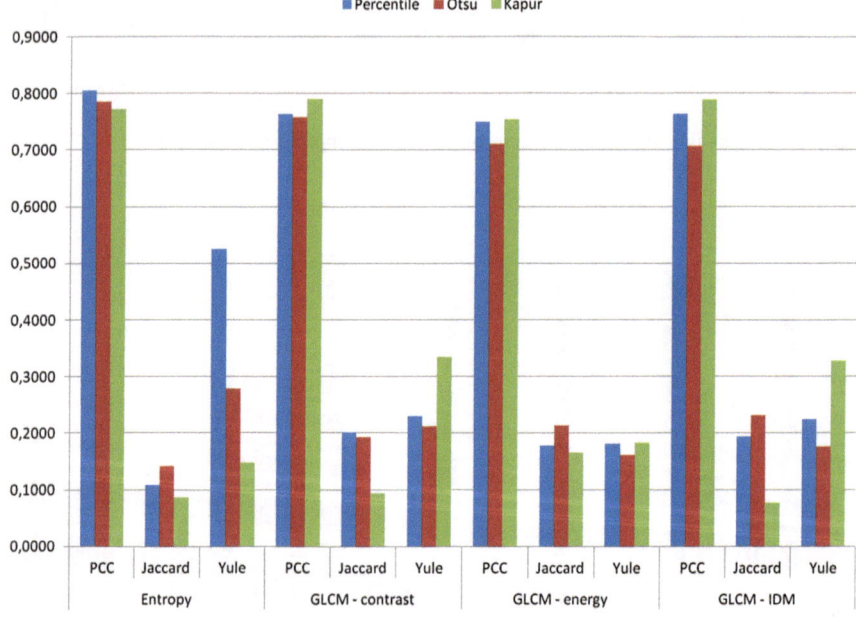

Fig. 7.7 Thresholding algorithms are compared in terms of texture-based change detection methods

For example, edge-based comparison produces thin lines representing object edges. Graph matching based method produces only points. Segment-based method measures the change in terms of segments. Thus, in structure-based change detection tests, we assume that the smallest element a method's output is an object. Objects are connected components which are set of spatially adjacent pixels. In order to handle object-based comparison, we adjusted our ground truth database in a way that the ground truth is represented by unconnected objects. To do so, we removed some parts such as roads and areas between structures which connect several objects. Based on this definition, the number of objects in all test image sets (from $AdanaP_1$ to $AdanaP_8$ and from $AnkaraP_1$ to $AnkaraP_{10}$) is 18, 10, 5, 14, 6, 9, 15, 14, 33, 75, 47, 46, 215, 113, 29, 21, 19, 28, respectively.

As we did for the pixel-based methods, we need to find comparison metrics between the ground truth and the result sets. We consider the following statistics before we define our performance criteria for object-based comparison. 'TP' is the number of truly detected objects in the ground truth image. Objects are assumed to be truly detected, if any object in the result image overlaps with a ground truth object. We also refer to these as matching objects. 'reTrue' is the number of objects in the result image which match an object in the ground truth image. 'FN' is the number of objects in the ground truth image which are labeled as negative by the method. 'reFalse' is the number of objects in the result image which do not overlap with any object in the ground truth image. Based on these measures, we benefit from two performance criteria named as Detection Percentage (DP) and Branching Factor (BF) used in [2]. These measures are

$$DP = \frac{TP}{TP + FN} \tag{7.8}$$

$$BF = \frac{reTrue}{reTrue + reFalse} \tag{7.9}$$

The detection percentage is the comparison of the number of truly detected objects to the number of ground truth objects. Branching factor decreases with respect to the increase in the amount of result image objects which do not match any ground truth object. Branching factor indicates noise objects in the result image.

A result image can obtain maximum detection percentage and branching factor by rendering the whole image as true. In this case, all ground truth objects are matched. Since there is only one object in the result image, 'reTrue' is equal to one and 'reFalse' is equal to zero. As a result, DP and BF will be equal to their maximum value of one. In order to prevent this case, we additionally define the Pixel Correspondence (PCorr) measure as

$$PCorr = \frac{rePxlOverlap}{rePxlTrue} \tag{7.10}$$

where 'rePxlOverlap' is the number of true pixels in the result image which overlap with an object in the ground truth image. 'rePxlTrue' is the number of true pixels in the result image. This measure gives an insight about the correspondence of the result and the ground truth images. If pixel correspondence is very low, other criteria are assumed to be unreliable. In Table 7.8 we provide the performance test results for the structure-based change detection methods.

As can be seen in Table 7.8, shadow-based comparison provides the best DP. Pixel correspondence cannot be used for interpreting shadow comparison results,

Table 7.8 Performance measures for the structure-based change detection methods

Structure type	TP	FN	reTrue	reFalse	DP	BF	Pcorr
Edge	314	403	419	643	0.4379	0.3945	0.2592
Matched filtering	475	242	1028	2387	0.6625	0.3010	0.3854
GMSR	326	391	615	1291	0.4547	0.3227	0.3674
Segment	183	534	213	382	0.2552	0.3580	0.3257
Local features	358	359	163	94	0.4993	0.6342	0.2980
Graph matching	405	312	2099	2882	0.5649	0.4214	0.4713
Shadow	521	196	765	1956	0.7266	0.2811	0.1717

Table 7.9 Performance measures for the pixel level fusion

	TP	TN	FP	FN	PCC	Jaccard	Yule
Fusion by binary AND	67371	2436223	34964	587228	0.8009	0.0977	0.4641
Fusion by binary OR	229651	2267787	203400	424948	0.7990	0.2677	0.3725
Fusion by association	195260	2338887	132300	459339	0.8107	0.2481	0.4320

because this method produces regions in the vicinity of the changed areas (not on them). When we discard PCorr for shadow comparison, it still has problems (low BF value indicates high noise in detection which also affects DP). Matched filtering based comparison produced the second best DP value. After that comes the graph matching-based comparison. When we compare these two, even though the matched filtering based method detected more changed objects than graph-based approach, it produced more noise. Graph matching based method has the second highest BF value after local feature-based comparison. Low PCorr of the local feature-based approach is because of the large areas produced by the sum of the Gaussians as we discussed in Sect. 5.5. This method produces few large components which have low association with the ground truth. Segment-based approach produced poor detection results. Other methods produced moderate results. Eventually, the graph matching based method produced high detection rate, has low noise, and has good association to the ground truth in pixel basis. Therefore, it is a good candidate for structural change detection.

7.2.6 Fusion of Change Detection Methods

We start with pixel-based category level fusion and provide the change detection results in Table 7.9. For comparison purposes, Table 7.4, which holds individual pixel-based change detection results, may also be used. As can be seen in Table 7.9, the association operator produced better PCC and Jaccard values compared to the individual pixel-based results while producing a comparable Yule value.

Table 7.10 Performance measures for the transformation level fusion

	TP	TN	FP	FN	PCC	Jaccard	Yule
Fusion by binary AND	1668	206763	240	44829	0.8222	0.0357	0.6960
Fusion by binary OR	15703	196987	10016	30794	0.8390	0.2779	0.4754
Fusion by association	14029	199926	7077	32468	0.8440	0.2619	0.5250

Table 7.11 Performance measures for the texture level fusion

	TP	TN	FP	FN	PCC	Jaccard	Yule
Fusion by binary AND	41048	2447327	23860	613551	0.7961	0.0605	0.4320
Fusion by binary OR	387319	1677810	793377	267280	0.6607	0.2675	0.1906
Fusion by association	272160	2060244	410943	382439	0.7462	0.2554	0.2419

Table 7.10 summarizes transformation based fusion results. To note here, poor performing methods (such as second, third, fourth principal components, and vegetation indices) are excluded from the transformation based fusion tests. Adding these methods to the fusion leads to labeling the whole region as either change or no-change. For individual transformation based change detection results, Table 7.5 may be of help. As can be seen in Table 7.10, fusion with binary association and binary OR operators produced good results especially in terms of the Jaccard coefficient. Binary AND fusion produced high Yule and low Jaccard values as expected.

The association operator again produced good results in texture-based fusion tests as given in Table 7.11. The Jaccard value of the association operator in texture-based fusion is better than all individual texture-based methods (tabulated in Table 7.6). Only the binary OR fusion operator produced a lower PCC value. In terms of the Yule coefficient, fusion by association again performed fairly well.

In structure-based fusion tests, we only have the association operator since the results can only be combined on an object basis. As we discussed in Sect. 7.2.5, the optimal performance was produced by taking the graph matching based method as base. We tabulate the result in Table 7.12. As can be seen, compared to the graph matching based method, association based fusion produced better DP and BF values with moderate pixel correspondence.

Inter-category level fusion tests are performed using the results obtained from category level fusion tests. For category level fusion, we only considered fusion by association results since it performed fairly well at every test. As the inter-category level fusion operator, we choose the association since structure category is involved in tests. Therefore, inter-category level fusion is the association of associations.

We first provide the fusion results in Table 7.13, where pixel and texture level fusion results were taken as base decisions. In these tests, results are in terms of PCC, Jaccard, and Yule measures since base decisions are pixel based. In this table, pixel association and texture association indicates the results taken from category level association fusion. Last two rows of the table provides the pixel structure and

Table 7.12 Performance measures for the structure level fusion

Structure Type	TP	FN	reTrue	reFalse	DP	BF	Pcorr
Edge	314	403	419	643	0.4379	0.3945	0.2592
Matched filtering	475	242	1028	2387	0.6625	0.3010	0.3854
GMSR	326	391	615	1291	0.4547	0.3227	0.3674
Segment	183	534	213	382	0.2552	0.3580	0.3257
Local features	358	359	163	94	0.4993	0.6342	0.2980
Graph matching	405	312	2099	2882	0.5649	0.4214	0.4713
Shadow	521	196	765	1956	0.7266	0.2811	0.1717
Fusion by association	466	251	305	283	0.6499	0.5187	0.3392

Table 7.13 Performance measures for pixel and texture structure inter-category level fusion

	TP	TN	FP	FN	PCC	Jaccard	Yule
Pixel association	195260	2338887	132300	459339	0.8107	0.2481	0.4320
Texture association	272160	2060244	410943	382439	0.7462	0.2554	0.2419
Pixel structure fusion	156363	2415050	56137	498236	0.8226	0.2200	0.5648
Texture structure fusion	236444	2200463	270724	418155	0.7796	0.2555	0.3065

Table 7.14 Performance measures for structure pixel and structure texture inter-category level fusion

	TP	FN	reTrue	reFalse	DP	BF	Pcorr
Structure association	466	251	305	283	0.6499	0.5187	0.3392
Structure pixel association	447	270	281	192	0.6234	0.5941	0.3487
Structure texture association	395	322	223	103	0.5509	0.6840	0.3716

texture structure fusion, where fusion result of the first category is taken as the base decision. These results indicate that, as we fuse structure with texture and pixel-based results separately, the performances improved.

We finally provide the results for the test where structure level fusion is taken as the base decision in Table 7.14. In this test, structure level fusion is associated with the pixel level and texture level fusion. As can be seen in this table, for both cases the DP value is decreased and the BF value is increased. This indicates an increase in the noise level after fusion by association.

References

1. Rosin, P.L., Ioannidis, E.: Evaluation of global image thresholding for change detection. Pattern Recognit. Lett. **24**(14), 2345–2356 (2003)
2. Lin, C., Nevatia, R.: Building detection and description from a single intensity image. Comput. Vis. Image Underst. **72**(2), 101–121 (1998)

Chapter 8
Final Comments

Abstract Two dimensional change detection methods are used extensively in image processing and remote sensing applications. In this study, we focused on these methods and their application to satellite images. We grouped change detection methods (based on the way they process data) under four categories as: pixel based, texture based, transformation based, and structural.

In pixel-based change detection, we explored several methods, such as image differencing, image ratioing, image regression, CVA, median filtering-based background subtraction, and pixelwise fuzzy XOR. The common characteristic of these methods is their being straightforward and cost efficient. Besides, they were able to detect changes as good as the other methods we explored.

Background subtraction-based change detection is well known in video image processing community. To our knowledge, this is the first time it is used in satellite images. This method has one shortcoming. It needs more than two images of the same area to provide meaningful results. Similarly, pixelwise fuzzy XOR-based change detection method is introduced in this study as a novel contribution to the community. One advantage of this method is that, it automatically provides a threshold value for change detection.

In transformation-based change detection, we explored several methods, such as PCA, KTT, vegetation index differencing, time-dependent vegetation indices, and color invariants. All of these methods depend on either color or multispectral information. For PCA, we observed that the first principal component is useful for change detection purposes. The greenness band of the KTT and vegetation indices were not useful for change detection in urban areas. On the other hand, the yellowness and brightness bands of the KTT and the TDVI performed fairly well for this purpose.

In texture-based change detection, we explored GLCM and entropy-based methods. Unfortunately, we could not get good results from them. One possible reason may be the texture descriptor. Although the used descriptors are accepted as a benchmark for texture analysis, they could not perform well in emphasizing the change.

M. İlsever and C. Ünsalan, *Two-Dimensional Change Detection Methods*, 71
SpringerBriefs in Computer Science, DOI: 10.1007/978-1-4471-4255-3_8,
© Cem Ünsalan 2012

For all these three change detection categories, we also considered threshold selection methods to automate the process. Therefore, we considered Otsu's thresholding method, Kapur's algorithm, and percentile-based thresholding. Our general observation is as follows. The percentile-based thresholding method tends to provide a lower threshold value compared to others. Besides, none of the three methods provided the best result.

Different from previous methods, we also considered structure information for change detection. In this category, we summarized the structure information based on edges, GMSR, matched filtering, mean shift segmentation, local features, graph formalism, and shadow. Based on these, we introduced a novel and generic change detection method. We were able to detect changes based on the structure extracted from two images as follows. If a structure (object) from one image overlaps with another object from the other image, this indicates no-change. The remaining objects indicate the change between these two images. We observed that, structure-based change detection methods are well suited for detecting changes in urban areas since in these man-made structure is dominant. To note here, structure-based methods only provide changed objects. They do not provide a pixel-based change map. We overcome this difficulty by using fusion methods. Among structure-based change detection methods, local feature and graph matching performed best. Unfortunately, they are computationally more complex compared to other methods.

Finally, we introduced fusion of change detection methods to improve the performance. Since different change detection methods summarize the change information in different ways, they can be fused to get a better performance. Therefore, we considered the decision level fusion based on binary logic AND and OR operations. We also developed a fusion method based on association. We applied fusion for both within category and between category levels. For both, we observed improvements in change detection accuracy.

As a conclusion, we can claim that 2D change detection methods considered in this study can be applied to detect changes in bi-temporal satellite images. They can also be used for change detection problems in general. The user can select the appropriate method for his or her needs.